Praise for Survive the Centu

"A sneaky-simple game that makes our climate choices real." – Brian Kahn, *Gizmodo*

"A climate fiction game where your choices on what narratives are highlighted will determine how well humanity will survive the 21st century." – Fridays for Future

"Laden with smart quips, tragic and hilarious narrative forks, and cute illustrations, and with its lengthy list credits (including artists, scientists, and backers), it's a lesson in what can be achieved through collaboration... you know, the sort we're going to need to keep average global temperatures down and our grandkids' odds of survival up." – Craig Wilson, *Input Mag*

"One thing I really liked about the game was how much of a holistic approach it took. You had to balance all of these different factors – climate, conflict, and the economy." – Forrest Brown, *Stories for Earth*

"The game prompts players to seriously contemplate challenging questions about how we can reach a balance between nature and ourselves." – *The Optimist Daily*

“Smart, even tragic at times, while offering an important lesson.” – Medha Dutta Yadav, *New Indian Express*

“Although the scenarios players encounter in the game are fictional — developed by a group of science-fiction writers — the creators collaborated with scientists to ensure the trajectories were possible within the bounds of time and warming.” – Katelyn Weisbrod, *Inside Climate News*

“For those who feel despondent and like it’s too late for us to do anything, go play *Survive the Century* and see that there are actually options.” – Tamara Oberholster, independent journalist

“A journey through the forking paths of policy choice and consequent scenarios juiced up with hilarious, satirical and irreverent stories.” – Rajat Chaudhuri, *Telegraph India*

A CLI-FI STORY OF CHOICE AND CONSEQUENCES

CREATED BY

Sam Beckbessinger, Christopher Trisos and Simon Nicholson

Survive the Century
survivethecentury.net

Published by Three Kids in a Trenchcoat

For all contributors, see the Credits (page 284)

ISBN (Print): 978-0-620-98748-6

Contents

How to play

You are the senior editor of the world's most popular and trusted news organization. You have the enviable power to set the news agenda, and thereby shift the zeitgeist.

That's right: YOU get to choose where this story goes.

Lead the world towards utopia, or unleash your inner villain to see how bad things can get. Nobody's judging. Your choices will determine how well humanity will survive the 21st century. No pressure.

Three quick things you need to know before we get started:

1. At the beginning of each new decade, you'll get a glimpse of what the headlines look like because of the decisions you've made. You can also read some news articles from the future written by prominent sci-fi writers. If you want to just read those stories, turn to page 270.
2. Sometimes, one of your choices will be up to chance. Have a six-sided die handy while you read this story (or type "roll dice" into Google).
3. At the end of this book, you'll find three trackers (page 291). You'll update them as you start each decade.

Set up your trackers, then turn to page 10 to make your first choice.

Oooh, we've got a rebel over here.

You COULD try just flipping through this like a normal boring book, but we can't promise that some of these pages won't be poisoned.

Don't say we didn't warn you.

To start playing TURN TO PAGE 10.

This page was soaked in poison.

Well, look. Now you're dead.

Turn to page 10 to start playing.

It's the early 2020s.

As an important and informed person, you're aware that the world is running out of time to address the climate emergency.

But the only thing your readers want to talk about is this pesky pandemic.

Rich countries are reaching nearly 100 per cent vaccination of their citizens, and are starting to open up again.

But poor countries, who haven't been able to afford vaccines, are seeing wave after wave of the virus. Experts are worried that it's continuing to mutate and to become more aggressive. They say our best chance is to get the whole world vaccinated.

What do you propose, oh, powerful editor of the world's most-read news site?

How should we get more vaccines to the rest of the world?

- Maybe a rich country could donate some vaccines? TURN TO PAGE 150.
- Maybe billionaires could donate some vaccines? TURN TO PAGE 58.
- Let's agree that each country should donate one per cent of its GDP to a global vaccine fund. TURN TO PAGE 86.
- Wake up, sheeple! Vaccines are a plot by sinister elites to turn everyone gay! TURN TO PAGE 184.

The world is rocked by a series of mass protests around inequality.

People demand change.

The movement is made up of disparate voices calling for different solutions.

Which of these solutions will your news site focus on?

- We need jobs! TURN TO PAGE 84.
- Let's just give everyone a universal basic income. TURN TO PAGE 152.

We deploy high-tech carbon removal technologies (literally giant machines that clear the air and ocean of global warming pollutants), and they're a real help in stabilizing the climate.

These machines are expensive, though. Activists point out that one way we could fund them would be to disband our militaries. The world has been a very peaceful place for decades, after all!

> *i*
>
> Carbon capture and storage technologies aim to remove carbon dioxide pollution from the air and store it where it can't re-enter the atmosphere, usually underground. Many of these technologies are still energy-intensive and expensive, and none are yet able to work at the large scales that would be needed to stabilize the climate.

Should we disband our militaries?

- Yes! Let's get rid of all the militaries. TURN TO PAGE 134.
- Hmm... maybe let's keep some kind of planetary defence system in place, just in case. TURN TO PAGE 97.

The News

All the news, all the time.

3 January 2030

UPDATE YOUR TRACKERS (PAGE 291)

ECONOMY **SWEET** TEMPERATURE **1.7** CONFLICT **SNIPPY**

TOP ARTICLE

Millions join protests as unemployment reaches record levels

Simultaneous protests in 57 countries over soaring income inequality

OTHER STORIES TODAY

Devastating floods in South Asia leave estimated 400,000 displaced

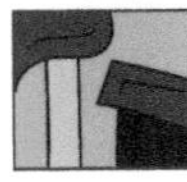

Coal keeping you cool? Record sales of air conditioners due to deadly heatwaves in areas of the tropics may just be making the problem worse, experts say

Anti-vax conspiracies dominate Reddit

Read more TURN TO PAGE 130.

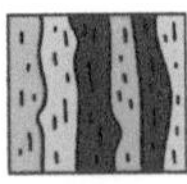

Report finds that the air is "unbreathable" in eight out of ten cities in China

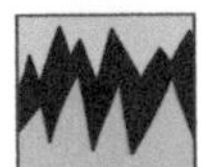

Environmentalists try to revive remaining polar bear populations after first ice-free Arctic summer

"The reefs are f-cked" admits leading coral reef expert

What will this new decade bring? **TURN TO PAGE 11.**

A poll indicates that an overwhelming 80 per cent of people believe that climate projects should be managed democratically.

A new generation of bold young politicians begin to speak out publicly about this issue. They try to pass laws nationalizing climate projects, or at least requiring them to follow some rules.

But after decades of pandering to the interests of private businesses, the government just doesn't have the power to force private companies to do anything anymore.

Corporations don't just run climate projects; they also control schools, private militaries, the healthcare system, and the entire financial system. They threaten to pull funding from climate projects completely if politicians keep pushing for nationalization.

Activists back down and nothing really changes.

Runaway effects of the climate hacks devastate parts of the world, forcing the last remaining free societies into corporate cities, where they hope to "earn" citizenship through work... maybe. One day. The unluckiest are forced into jobs on the nascent Mars colony.

- You made it to 2080! TURN TO PAGE 235.

UPDATE YOUR TRACKERS (PAGE 291)

ECONOMY **NO CHANGE** TEMPERATURE **NO CHANGE** CONFLICT **NO CHANGE**

PRESS RELEASE

Trouble in Paradys: Virtual worlds, real targets

After a series of high-profile burglaries in an exclusive paradise zone, VR critics wonder: How real is too real?

BY LAUREN BEUKES

→

Paradys is two things: a real-world paradise zone for the rich and famous who are not quite rich and famous enough to get to Mars, and the most popular virtual world where your avatar can rub shoulders with a who's who of 21st-century celebrities from Beyoncé and Jay-Z for the nostalgiaphiles to breakout VR influencers like Star*Star and M@keMi. ParadysVR is known for its attention to detail, exactly replicating the beachside mansions and vineyard villas and penthouse apartments of its real-world tenants that you've seen on your entertainment network of choice, promising to catapult you into the most ReallyReal™ Rich & Famous experience possible.

Unfortunately, it's exactly that high fidelity to life that has allowed an elite gang of criminals to precisely home in on their targets, using the virtual world to map the layout and exploit security weak points.

Paradys is currently down for maintenance while ReallyReal™ restructures, stranding millions of people in their real lives – a move that's "shockingly inhumane", according to one human-rights watchdog, VR4All. Director Allison Engels says they will be suing ReallyReal™ for damages.

"We're talking about 5.5 billion climate refugees living in the worst-possible conditions. Virtual reality is vital for their mental health, for them to be able to have a sense of purpose and value, to be able to connect with friends and family, living or dead, through their avatars. In the two days it's been down, we've already seen an uptick in suicides and domestic violence."

→

ReallyReal™ spokesperson, Colin Ray, says, "It's regrettable that a few bad apples would ruin it for the majority of our customer base, who are ordinary honest citizens desiring the luxury of the perfect escape. But we're convinced that our new proprietary obfuscation technology that randomizes key details in every load-up will have us back up and running in no time."

Investigations are ongoing and police say although the gang may have used VR to scout out the locations, they believe there was someone on the inside who let them into the real-world Paradys and helped facilitate the burglaries. Paradys residents with any information are urged to come forward.

You made it to the end of the century! **TURN TO PAGE 109.**

Hey, this could be a chance for us to fix the economy and address the climate crisis at the same time!

We choose to pass sweeping "Green New Deal"-style plans that create a lot of jobs building renewable energy plants and efficient public infrastructure like trains, and making existing buildings more energy efficient.

A lot of these projects are community-run and community-owned, like local solar panel installation networks.

This makes communities more resilient to extreme weather, raises the standard of living for the less well-off, and accelerates the transition away from dirty energy like coal, oil and gas.

A consortium of oil companies wants to buy a lot of advertising space, arguing that we're moving too fast on reducing emissions and this is going to damage the economy.

i

The Green New Deal is a plan put forward by New York Representative Alexandria Ocasio-Cortez for tackling climate change. It calls on the US government to wean itself off fossil fuels and to invest in high-paying jobs in clean-energy industries that can simultaneously address other types of social problems like economic inequality, racial injustice and gender discrimination. An example would be a country investing in new high-speed train networks that

→

would both reduce the number of cars on the road and create new jobs. Other countries have proposed similar plans.

Will you let them?

- Sure, you can use the extra cash to buy new sweatpants. TURN TO PAGE 208.
- No. In fact, let's publish something about how we're not moving *fast enough*. TURN TO PAGE 177.

The world is falling far short of the (pitiful) climate change targets agreed to, and the worst impacts are happening in poor countries.

The leaders of 50 countries in the Global South (low-income countries) band together and say *enough*. This is our last shot. We have to stop all emissions within the next 15 years, and urgently try to reverse the damage that's already been done.

Most of your readers live in rich countries. Will you encourage them to take responsibility and get serious about reversing the damage we've done to the planet?

- Absolutely. The rich world should accept full responsibility and pay for the damage that's been done. TURN TO PAGE 37.
- The rich world should absolutely stop polluting, but every country needs to find its own solutions. TURN TO PAGE 64.
- Rude! Those upstarts need to be put in their place. TURN TO PAGE 40.
- Surely some cheap magical technology could just fix this? TURN TO PAGE 193.

UPDATE YOUR TRACKERS (PAGE 291)

ECONOMY **NO CHANGE** TEMPERATURE **NO CHANGE** CONFLICT **NO CHANGE**

PRESS RELEASE

Review of a "Green Mall" by a grumpy Gen X shopaholic

BY MARIA TURTSCHANINOFF

I went to this mall with my grandchild, who assured me it was "dope". I don't like it when the youth of today assume the sayings of my generation. Get off my lawn, kids! Speaking of lawns, there were way too →

many of them at this mall. A mall should be inside a huge concrete building, and lawns should be well clipped and outside my house. Not on top of roofs! On top of that, these lawns had probably never seen a mower and were full of wildflowers and butterflies and bugs and things. NOT sanitary. I think I even saw a goat grazing? In a city? Jeez.

I deduct the first star because a bird pooped on me when we got there. This is why we need to stop biking everywhere and resume driving in cars. I get way too much fresh air and bird poop.

But the worst thing about this mall was that they didn't sell any stuff! I want to go to a mall and be mesmerized by all the shiny things that I can't afford and then buy them anyway and cry when my credit card bill comes. This mall consisted of small buildings connected with walkways lined with trees (more damn birds), and in each building there was a repair shop. Repair shops for phones and washers and pots and clothes and shoes – wait, I guess those are called cobblers – and electrical things and what have you. I know my grandparents used to fix things when they broke, but not my generation! Our things weren't built to last. We would toss it out and buy something new. A new phone every year, new clothes weekly and new furniture whenever the fancy struck. OK, I know not all of the world lived like this. But everyone I knew sure did.

I blame this recycling and re-using on those damn Gen Alphas. They care too much. Now my generation, we knew how not to give a shit. Sure, we'd recycle the odd bottle and try to turn off lights when we left →

the room. But this generation – they like actually want to make a difference. Maybe because the world is hotter than it's been in like, ever? Who knows. Not that they're nearly as bad as my grandkids' generation. Gen Green. Ick. What a name, am I right?

I didn't get to buy a single item, and that's another star off this review.

I heard the person who came up with the concept of the repair mall would be rich now, if people still got rich as they did back in my day. Now she's just a global hero or something. Whatever. The mall had a food court which was almost like the ones in my day, except with no disposable cutlery or plates. The food was quite good, even though it was all locally sourced. They had a blueberry cobbler that tasted like my grandma's. I wept when I ate it. I don't like crying in public, which is why I will deduct the last star I was going to give.

Sincerely,
A Gen X shopaholic

Let's see what the 2050s will bring. **TURN TO PAGE 50.**

Doesn't it make more sense for the government to just give people cash and let them spend it how they want to?

Governments tend to be less efficient than specialists. So we should leave education to the education experts, and healthcare to the healthcare experts!

We dramatically defund healthcare, education, energy and other social services and lower taxes, instead giving every citizen cash every month. There's a boom in new businesses that pop up to offer these services, but the quality is variable.

Award yourself the UNIVERSAL BASIC INCOME badge (page 289)!

Climate change is still a worry, but governments don't have a lot of money left to fix the problem directly.

- Instead, they offer grants to tech companies who are working on solutions.
TURN TO PAGE 52.

Billionaires make a compelling argument:

In return for ownership of all of this lovely wild paradise land, they will invest huge amounts of money fixing the climate once and for all.

- Oh, wow. Thanks, billionaires! TURN TO PAGE 38.

Pressure mounts on national leaders.

They realise that if they don't make some changes, they might lose the next election.

- This leads to the calling of a new emergency international climate summit.

TURN TO PAGE 98.

So, where do we end up by the end of the century?

We will look back on this as one of the bleakest times in human history.

Like the fall of Rome or Europe's Dark Ages, a reminder that humanity can go backwards as well as forwards. The gains of the 20th century have mostly been wiped out. Average lifespans have returned to pre-1900s levels. We will rebuild the world, but so much has been lost.

How did you get here?
You allowed climate change to be a source of conflict between people who saw everything as a zero-sum game.

To take action **TURN TO PAGE 268.**

Play with the data behind your decisions at tinyurl.com/4xe9jz6d

The Mars colony is going to cost an eye-melting amount of money.

Award yourself the I WENT TO MARS AND ALL I GOT WAS THIS LOUSY BADGE badge (page 289)!

Most geoengineering projects are defunded, and many of them have to be stopped suddenly.

It turns out that we'd become more dependent on geoengineering than we realized. As soon as we stop, the temperature shoots up very suddenly.

i

One of the potential risks of blocking some of the sun's rays to offset global warming is that if we stopped, the temperature could rise very quickly. This would happen if geoengineering was deployed without society also reducing greenhouse gas emissions at the same time. Suddenly stopping geoengineering would cause temperatures to rebound fast. Scientists call this risk "termination shock" and it would be catastrophic for ecosystems globally.

- This is BAD. TURN TO PAGE 221.

You get a frantic phone call in the middle of the night:

An elite Chinese military operation has knocked out America's sulphur dioxide pipe.

Your lead climate reporter tells you, in a voice choked with fear, that this means the world is going to warm as much over the next two years as it has in the last two decades.

- Sorry, you're all out of choices. TURN TO PAGE 49.

The losers of the climate wars continue to try to rebuild.

Desperate, their economies in tatters, a lot of them see small doomsday cults emerge, some of them with nuclear bombs.

The leader of one of these cults wants to publish an open letter on your news site arguing it would be better to just drop nukes and end everything.

Are you going to let her publish her letter?

- Yes, let's end it all. TURN TO PAGE 128.
- No, it's not over yet. TURN TO PAGE 207.

UPDATE YOUR TRACKERS (PAGE 291)
ECONOMY **NO CHANGE** TEMPERATURE **NO CHANGE** CONFLICT **NO CHANGE**

PRESS RELEASE

Quiz: What's your ticket to Mars?

BY LAUREN BEUKES

Ah, Mars, that russet interstellar paradise we mere mortals can only gaze up at in envy, through the blur of smog and light pollution and the white glare of the sulphur dioxide skein in the atmosphere. It's a dream come true among the terraformed lava tubes, if you could only get there. Mars is a paradise for the lucky few who can afford it, but there are other ways to get a piece of that red life. Take our quiz to find out what Mars-specific job you'd be best suited for.

→

Pick a romantic Mars vacation spot:

A. A rustic cabin on the ski slopes of Olympus Mons outside Lovelace One
B. At the Hilton Honeycomb in Medusa Fossa, where the rooms have been carved from the lava tubes to resemble the strange rock formations
C. At Sun International's Tropica, an immaculate rain forest family resort with luxury cabanas, waterslides, nightlife and shopping
D. A private Tuscan-style villa on the shore of the newly-filled-in Valles Marineris dam
E. In a radiation-proof tent at the South Pole - bring on the adventure!

Pick a horror movie monster that genuinely frightens you:

A. The killer clown
B. Bodysnatching aliens
C. Serial killer
D. Zombies
E. Evil dolls

→

If you had to stay on Earth, which corporate haven would you choose?

A. Gatesville
B. HSBC-Hariri
C. The Republic of Jeff
D. Kanye Island (Ukraine)
E. Teslamerica

Choose one essential to take with you:

A. A family keepsake
B. A camera
C. Your mom
D. Your best party outfit
E. A multitool

You get a Mars-appropriate genetically modified pet! What's it going to be?

A. A low-gravity dog
B. A really exotic melange like a tigerslothtardigrade
C. No animals, I'm allergic
D. A racehorse
E. Something I can eat if the going gets tough

→

Pick your Earth-favourite comfort food to eat on the long journey there:

A. Mac and cheese
B. Intravenous nutrition infusions – I want the full cosmonaut experience, please
C. Dhal curry and grilled vegetables
D. Real-fish sushi; it's probably going to be the last time
E. Barbeque – I want all that sweet protein

Finally, what's your hidden talent?

A. A robust immune system
B. Unique artistic vision
C. I'm nurturing
D. I'm hot, sophisticated and charming AF
E. Manipulating people

Mostly As: You're healthy, physically fit, down to earth, practical, and definitely not the squeamish type. You're perfect for a COHO, a concentrated organ-harvesting operation, which sounds a lot scarier than it is. You'll be growing extra organs for transplant alongside your own, which can be harvested without (much) risk. But don't think growing an extra liver means you can drink double the amount of alcohol. Because of the lower oxygen levels on Mars, you get drunk a lot →

quicker – and there's the medical compact where you agree not to take any intoxicants during the growth period. You may be asked to trial new vaccines and medicines, though, which will put you on the cutting edge of scientific developments and saving rich people's lives!

Mostly Bs: It turns out AI can make killer pop hits and write accessible thrillers, but it can't (yet) create art that is surprising and profound on a human level. If you've got a special talent and a truly unique voice, you can apply to one of the Red Residencies which fly out performers, musicians, dancers, artists, sculptors, slam poets, and the occasional novelist to be inspired by life on Mars for anywhere from six months to two years. The truly exceptional may be granted permanent residency under the De Medici patronage programme, but you better be able to keep it fresh and stay ahead of artificial intelligence.

Mostly Cs: You're warm and nurturing, healthy and of sound mind, and you don't mind being pampered or told what to do. Hopefully you're also still in your 20s and with a functional and healthy womb. Pregnancy surrogates are in hot demand on Mars, especially for older couples or women who don't want to go through the rigours of incubating a human on a hostile planet and the toll it takes on your body, which can be severe. Most surrogates only last one pregnancy or possibly two before they're shipped back to Earth, but there have been some special cases

→

where the couple in question kept the surrogate on to be a live-in nanny.

Mostly Ds: You have a taste for the high life and your sights set on the stars. It helps that you're young, beautiful, urbane and incredibly charismatic. You're the ideal trophy spouse for a billionaire – if you can snag their eye. After all, there's a lot of competition.

Mostly Es: You're a hustler and a charmer and a schemer, able to outwit, outlast and outplay. Maybe you already made it to the final four on one of the Earth-bound seasons of *Survivor*, but this time you're playing for real, for lifetime rights to a coveted spot in a Mars paradise zone.

Onwards into the 2070s! **TURN TO PAGE 165.**

Each rich country agrees to contribute ten per cent of its GDP to a global climate transition fund.

This money is spent to build infrastructure to help cities ravaged by climate change adapt to extreme weather, to invest in food security, and to pay compensation to countries that have already suffered the most from global warming. We also invest in researching technological ways that could reduce global warming.

Most countries that did not already offer a universal basic income (UBI) now introduce them.

Quality of life improves rapidly for most people in the world.

If you don't already have one, award yourself the UNIVERSAL BASIC INCOME badge (page 289)!

- You made it to 2050. TURN TO PAGE 59.

We decide that the world's best chance of fixing the climate with technology is if private businesses do it.

Our billionaire playboy overlords start deploying huge, flashy technology projects (all named after themselves) to re-engineer the climate.

Overall, the global temperature does start to drop, but there are some unfortunate side effects. Weather patterns are disrupted, leading to awful droughts in some parts of the world, and floods in others. Mexico City, Cape Town and Tokyo all have to introduce water rationing, while Beijing and New York suffer the worst flooding in their histories.

We also send up the first manned Mars missions, so we've got that as a backup. Award yourself the I WENT TO MARS AND ALL I GOT WAS THIS LOUSY BADGE badge (page 289)!

i

Billions of dollars have been spent on Mars exploration. Some estimates suggest that bringing the first four people to Mars could cost $6 billion. But many scientists, like Lucianne Walkowicz who works for NASA, say we should stop looking for a backup planet in case climate change makes Earth uninhabitable. Instead, she argues, we need to refocus our energy on preserving our planet.

- Woohoo! You made it to 2060. TURN TO PAGE 149.

Improvements in the space mirror fleet and other adaptation technologies help, but a lot of the outside world is still a hostile, frightening place.

Most urban people retreat inside climate-controlled domes, or build cities underground.

The idea of "nature" or "wild space" becomes a distant memory.

Some crazy fringe religious groups hate the domes and the decadent lifestyles that take place in there. And some of these groups have bioweapons.

Should we quash these terrorist groups?

- Yes. TURN TO PAGE 198.
- No. TURN TO PAGE 260.

The News
All the news, all the time.

3 January 2050

UPDATE YOUR TRACKERS (PAGE 291)

ECONOMY **HMMM...** TEMPERATURE **2.25** CONFLICT **DANGER**

TOP ARTICLE

China to re-home entire population of Bengal tigers after flooding leaves region uninhabitable

Bangladeshi humans not as lucky: Millions of displaced people wait at their border for entry permits

OTHER STORIES TODAY

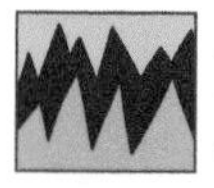

Advanced countries reap benefits as brain drain peaks, leaving towns empty in the developing world

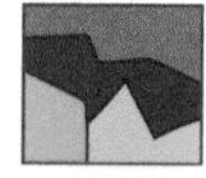

US puts refugees to work in "mine towns" in a massive carbon dioxide removal project despite humanitarian concerns

Mass abductions hobble India's guest talent programme

Read more **TURN TO PAGE 181.**

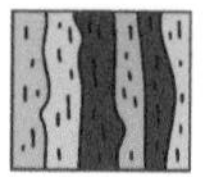

Kenya's Serengeti-Mara National Park is de-proclaimed as global food insecurity increases the demand for and value of land with agricultural potential

Let's see what the 2050s will bring. **TURN TO PAGE 161.**

You publish an editorial arguing that we need to try a geo-engineering solution. But what will that solution look like?

Roll a die!

- If you roll a 1, TURN TO PAGE 80.
- If you roll a 2 or 3, TURN TO PAGE 38.
- If you roll a 4, TURN TO PAGE 65.
- If you roll a 5 or 6, TURN TO PAGE 217.

The world launches a global climate fund: a taxation pool that taxes those who produce pollution or consume luxury goods, and uses this money to fund these projects to rewild and protect the environment.

Global South countries band together and argue that they shouldn't have to contribute to this pool, and also that old IMF loans should be forgiven, because of the historical debt of colonialism.

Should rich countries agree to this?

- That seems fair. TURN TO PAGE 195.
- Absolutely not! TURN TO PAGE 218.

UPDATE YOUR TRACKERS (PAGE 291)
ECONOMY **NO CHANGE** TEMPERATURE **NO CHANGE** CONFLICT **NO CHANGE**

PRESS RELEASE

Matrimonial advertisement: Indian NRI seeks suitable match

Thanks to our reporter Rajat Chaudhuri for finding this curious advertisement on a local classifieds site.

BY RAJAT CHAUDHURI

Hi, this is Chandan. Indian, non-resident Indian (NRI), clean-shaven male of 27 years with a PhD in sports medicine, seeking a suitable alliance with Indian Hindu cisgender non-veg woman – caste no bar. I hold a stable job with decent pay and climate-insured accommodation in a Chinese MNC-run gambling city 50 kilometres from Johannesburg, South Africa. My employment conditions offer reasonably excellent labour rights with handsome perks including discount-price access to Chinese super-vaccine, Coronacraft, which guarantees protection against all future mutations of novel coronaviruses. I am looking for a strong-willed and like-minded partner of milk-white complexion, with no siblings.

→

My current location offers robust food security under the US embassy's McBush animal-protein programme for 12 months supplemented by Chinese Communist Party-certified TruSafe GM rice, both renewable annually, subject to good-behaviour criteria. My ideal partner would be someone with low bodyweight and a healthy appetite for fun.

I am honest, hardworking, and believe in reincarnation. My role at the company is that of stretcher-bearer, first class. Low-risk job, involves the gentle bearing of the most esteemed guests of probability theatres (new nomenclature for gambling establishments) when they are unable to move by virtue of excitement, shock or disagreements with security personnel.

With gold-standard morals, I have accumulated good karma and look forward to sharing it with a life partner who has all requisite vaccinations for the updated zoonotic disease outbreak list of the WHO. My only weakness is a fear of spectres, whose numbers have exploded after a year of flash floods and raging wildfires. I look forward to offset this dread with the nectar of conjugal bliss.

Looking for a spouse, from a good educational background, who doesn't harbour any inclination to generate offspring, as this tends to destabilize the current economic scenario and is a weight on the planet. Instead I foresee a child-free union, which will augment our store of good karma, granting us higher stations in our next life – in a hydrogen-powered future.

→

Please connect via IM with your bodyweight and enrolment number.

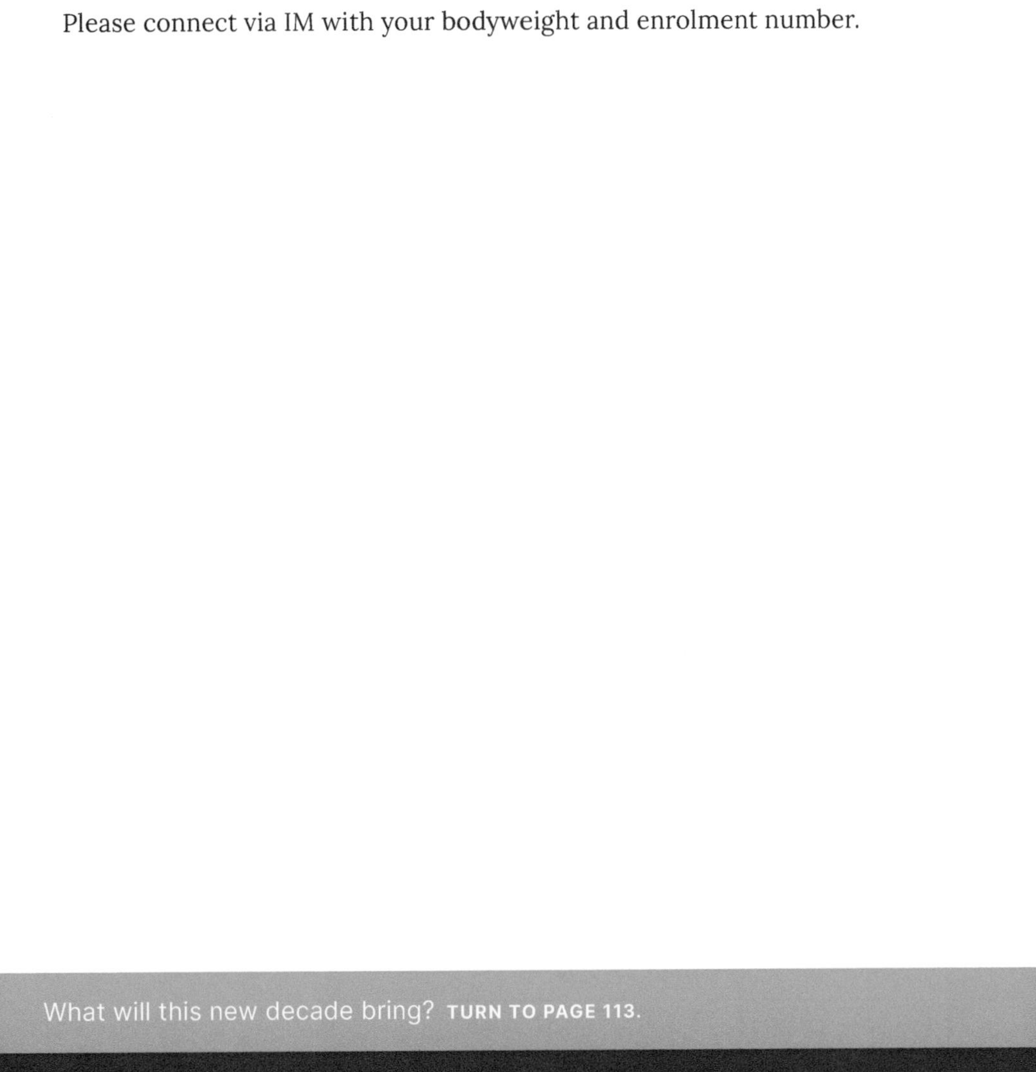

What will this new decade bring? **TURN TO PAGE 113.**

You publish an editorial encouraging anyone to try a last-minute climate hack.

Roll a die!

- If you roll a 1, TURN TO PAGE 80.
- If you roll a 2 or 3, TURN TO PAGE 139.
- If you roll a 4 or higher, TURN TO PAGE 232.

We decide it's probably more important to terraform the planet we're already on.

We invest more money into climate-hacking projects, with mixed success but no civilization-ending disasters.

There's still a Mars colony, but it's small, the passion project of a single billionaire.

- Well, you made it through the 2060s. TURN TO PAGE 231.

We tax the rich and invest in public infrastructure and generous grant programmes.

Award yourself the UNIVERSAL BASIC INCOME badge (page 289)!

We start to see a lot of experiments in new forms of communal and sustainable urban living, as people are freed from constant worry about their next pay cheque.

We start reforming capitalism with new rules like employee representation on boards. The gap between rich and poor closes.

A lot of renewable energy projects happen on a grassroots level as co-operatives and community-owned projects. It looks like the world is on track to reach zero emissions by 2037.

But wait, the rich have #sadfeels about all this! They produce charts proving that all this equality will decimate the world's economy!

- Well then, maybe we're measuring the economy wrong. TURN TO PAGE 76.
- Oh no! Better undo all of these reforms. Things were fine before, weren't they? TURN TO PAGE 155.

The effects are terrible: mass extinctions, devastated crops.

The biggest impacts are in the oceans – in the first year, fish stocks halve. Then they halve again.

The eastern allies led by the United States finally surrender. The war is over.

Millions starve. Billions are displaced. This decade is called "the terrible 70s".

Award yourself the I SURVIVED THE 70s badge (page 289).

- When will this decade end? TURN TO PAGE 111.

There's all this new wild space in different parts of the world.

Species have more space to adapt to climate change and some are recovering. Nature is beautiful and valuable. People would love to get their hands on it. We could use this for tourism! Or let rich people live there! Or exploit newly discovered resources! Corporations also want to fund (and manage) them because they're carbon sinks, to offset their activity elsewhere.

Award yourself the I SAVED THE ORANGUTANS badge (page 289)!

What should we do with all this lovely wild space?

- Fund indigenous communities and small co-operatives to keep running them. TURN TO PAGE 241.
- Let rich people build private paradises there. TURN TO PAGE 25.
- Set up central committees of ecological specialists responsible for managing the wild. TURN TO PAGE 42.

Alternative forms of society start to emerge in the hinterlands.

They're tough landscapes, but these become incubators for alternative economies and value systems.

Also Thunderdomes.

- You made it to 2050. TURN TO PAGE 146.

We launch a number of green-technology funds that invest in private businesses researching new ways to reduce climate change, or help the world adapt to it.

This sparks a lot of innovation, and mints a whole new generation of tech billionaires and green influencers. Green is the new gold, baby!

- Congrats! You made it to 2040. TURN TO PAGE 99.

We stimulate economies by selling new coal-mining/drilling/ fracking rights.

The owners in these industries get rich and national coffers get a temporary boost.

Pollution gets much worse. People in cities powered by coal and congested with cars say they're struggling to breathe, and their children are falling sick.

i

The World Health Organization estimates that 4.2 million people die every year because of air pollution (as of 2021). Reducing the amount of pollution in the air by switching to renewable energy not only reduces the impact of climate change in the future, but also dramatically improves the health of people around the world today.

- You made it to 2030. TURN TO PAGE 13.

We choose to rewild as much of the world as we can.

Huge areas of land are now protected natural spaces given back to indigenous communities to steward, or managed by non-profit organizations.

Restoring so much land to its natural state allows for many formerly threatened species to start thriving again. Natural forests clean the air and help to trap greenhouse gases, speeding up our efforts to curb climate change.

Award yourself the I SAVED THE ORANGUTANS badge (page 289)!

- You made it to 2050. TURN TO PAGE 175.

The News
All the news, all the time.

3 January 2040

UPDATE YOUR TRACKERS (PAGE 291)

ECONOMY **ROCKING!** TEMPERATURE **2.2** CONFLICT **FRIENDLY**

TOP ARTICLE

“Birds are just dropping dead from the trees,” says tearful PM

Record-breaking heatwave causes mass animal deaths across Australia

OTHER STORIES TODAY

Millions of US citizens move north to escape heat and rising storm surges

Divorce rates jump after universal basic income introduced in more countries: “Good riddance!” say millions of thrilled newly single women exiting bad marriages

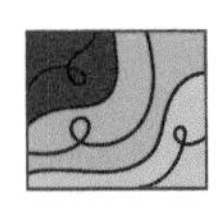

Choosing a beautiful air-pollution mask to match your wedding dress: Tips from our style experts

Dubai to host world’s first indoor winter Olympics

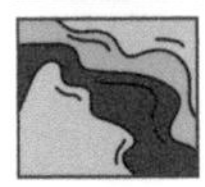

Reports of falling fish stocks in tropics: “Millions could starve,” warn food-security experts

South Korean inventor debuts “floating farms” that can migrate with the weather

People are feeling optimistic about the 2040s! **TURN TO PAGE 159.**

We hold another global geoengineering summit.

More leaders are willing to try geoengineering in theory, but there's still some debate about how exactly all this would be managed and who would benefit.

- Hang on, we're getting emergency news out of the US about a new virus...
  TURN TO PAGE 221.

Each new strongman leader blames the others for creating the migrant crisis.

- International tensions are heating up. TURN TO PAGE 210.

The richest people in the world band together and donate vaccines to countries that need them.

People erect statues in honour of their billionaire saviours.

The rollout programmes aren't very effective, though, because they're optimized for PR rather than public health. Instead of taking months to roll out the vaccine, it takes two years.

By the late 2020s, economies are battered. Countries need bold new economic-recovery plans. Different groups propose different solutions.

Which economic proposal will your news site give the most time to?

- Governments should invest in large-scale public infrastructure projects, like trains and housing. TURN TO PAGE 18.
- Cut taxes and regulations on businesses so they've got more money to grow. TURN TO PAGE 148.
- Our country first! Let's break with our international trade agreements, put tariffs on imports and stimulate our own economy. TURN TO PAGE 224.

The News
All the news, all the time.

3 January 2050

UPDATE YOUR TRACKERS (PAGE 291)

ECONOMY **ROCKING!** TEMPERATURE **3** CONFLICT **FRIENDLY**

TOP ARTICLE

Over 50 new zoonotic viruses detected this year as warming moves previously isolated animal species into contact with each other

No big deal, say health experts, because global health surveillance system is so effective

OTHER STORIES TODAY

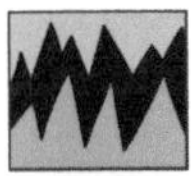

Half of all insect and plant species heading for extinction, say experts

Interest in VR safari holidays at all-time high

Amazon forced to replace 90 per cent of workforce with drones after UBI leads to mass walkout of low-paid employees

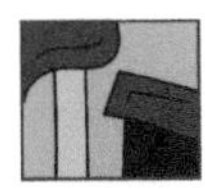

Versace premieres "cooling suits" at Paris Fashion Week allowing fashionistas to go outside in the tropics

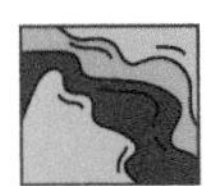

Russia and Canada reach agreement over who owns the new year-long sailing routes over the ice-free Arctic Ocean

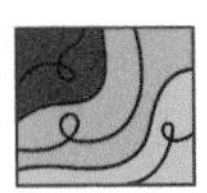

Japan becomes first country to be powered entirely by hydrogen

Let's see what the 2050s will bring. **TURN TO PAGE 92.**

We tax the rich and invest in public infrastructure and generous grant programmes.

We start to see a lot of experiments in new forms of communal and sustainable urban living, as people are freed from constant worry about their next pay cheque.

We start reforming capitalism with new rules like employee representation on boards. The gap between rich and poor closes.

A lot of the renewable energy projects are run as co-operatives and community-owned projects, and the technology is shared freely.

The rich have #sadfeels about all this. Nobody cares.

- Congrats! You made it to 2040! TURN TO PAGE 219.

Nation states do less and less – they're just an administrative hassle and a barrier to truly fair global decision-making.

We finally abolish nation states and replace them with a network of local democracies, with fair representation from every community in the world on a central world council. Citizens are very active within the politics of their own cities or small communities.

- Everything's going great! TURN TO PAGE 141.

America's rogue geoengineering is the last straw.

This sparks a decade-long world war.

There are geoengineering and counter-geoengineering attacks on both sides, and vicious proxy wars all around the world. Millions of people die, and the world's economy is devastated.

After nine years, there's still no clear winner. People are exhausted of fighting. The leaders of both sides agree to a clandestine meeting.

What do you think, editor? Should they agree to a truce?

- Never surrender! TURN TO PAGE 240.
- Yes, heck, this has gone on long enough. TURN TO PAGE 74.

The News

All the news, all the time.

9 August 2090

UPDATE YOUR TRACKERS (PAGE 291)

ECONOMY **OH NO...** TEMPERATURE **3.7** CONFLICT **DANGER!**

TOP ARTICLE

Indian tour operators advertise winter scuba tours of the sunken city of Malé

Booming disaster-tourism industry turns pain into profit

OTHER STORIES TODAY

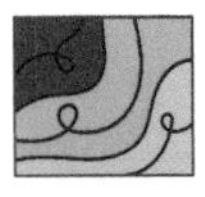

Food scientist warns that synthetic rice, beans may not be as safe as manufacturers claim

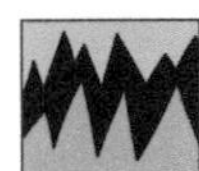

Two floating cities nearly collide off coast of Madagascar

Thousands complain as heatwaves cripple mobile-phone use outdoors, teens mock Gen Z for not using refrigerated phone cases

Bats drive pigeons out of Kingston, Jamaica, take over kudzu-choked remains of the sunken city

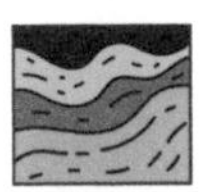

Last holdouts finally abandon the Sahel

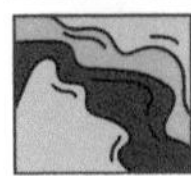

Global fish populations collapse due to ocean acidification

You made it to the end of the century! **TURN TO PAGE 172.**

Rich countries agree to reduce their own emissions, but refuse to contribute to a global climate change fund.

“Let every country deal with their own problems,” they say.

In polluting countries, this leads to a patchwork of ineffective solutions like a personal sustainability score to try to get individuals to change their consumption habits, and some agreements allow businesses to buy and trade pollution credits. Some cities are now fully “green”, but others are emitting more than ever.

- Oh, but hold up! You’re getting word of some big news in Greenland. TURN TO PAGE 220.

A global agreement seems impossible.

Years of drought and crop failures lead William Trump, America's newly appointed Leader-for-Life, to erect a huge vertical pipe spraying reflective sulphur dioxide into the upper atmosphere in a last-ditch attempt to cool the planet (yes, there are a lot of dick jokes).

This cools America but causes drought across India and southern Africa.

- What?! It's 2060 ALREADY?? TURN TO PAGE 75.

Oh, whoops, we did not invest in a planetary clean-up project.

It just didn't seem necessary!

Without the space mirrors keeping the planet cool, things heat up FAST. The world experiences the hottest year ever. Most major coastal cities are flooded. People die of heatstroke. Wild animals from warm parts of the world have to be ushered into temporary air-conditioned enclosures to keep them alive.

Most of the remaining insects die out, including all bees. The Council scrambles to figure out a solution before everyone dies of hunger.

Award yourself the OH, WERE THESE IMPORTANT badge (page 289)!

Roll a die!

i

One of the potential risks of blocking some of the sun's rays to offset global warming is that if we stopped, the temperature could rise very quickly. This would happen if geoengineering was deployed without society also reducing greenhouse gas emissions at the same time. Suddenly stopping geoengineering would cause temperatures to rebound fast. Scientists call this risk "termination shock" and it would be catastrophic for ecosystems globally.

→

Will our magnificent Council figure something out in time?

- If you roll 2 or less, TURN TO PAGE 120.
- If you roll higher than 2, TURN TO PAGE 187.

UPDATE YOUR TRACKERS (PAGE 291)
ECONOMY **NO CHANGE** TEMPERATURE **NO CHANGE** CONFLICT **NO CHANGE**

PRESS RELEASE

Report from the Global Indigenous Peoples' Stewardship Summit

BY MARIA TURTSCHANINOFF

"When we began the Global Indigenous Peoples' Stewardship (GIPS) programme, many non-indigenous people thought this meant that everyone had to live in tents or huts in the wilderness," elder Ásllat-Mihku Ilmara Mika Petra, or Petra Laiti as she is also known, says with a wry smile. "Now look at the world: the floating botanical cities of Korea, the underwater city of Oceania in Japan, and Green New York."

Laiti has been showing me around the site for the GIPSS in Rásttigáisá, Sápmi. The site sits at the foot of Mount Rásttigáisá.

"Our founding summit was in Australia," Laiti says as we sit down in

→

one of the communal teepees, or kotas as they are known in Sámi. "This time it was thought appropriate to have the summit in the northern hemisphere." A young woman in what I later learn is the traditional garb of the Anishinaabe people of Canada serves us coffee. During my stay, I have seen people in both Western clothes and traditional garb. But the people from the southern hemisphere tend to either adopt the clothing customs of their northern sisters and brothers or wear Western clothes – their traditional garb usually is too cold.

It is June, which means the sun never sets, the mosquitoes are plentiful and the days can be pleasant, but never very hot.

"This was a benefit during the height of the climate change," tells Aslak Lihtonen, a steward elder from the western region of Sápmi. "We had a lot of climate refugees here then. It was a difficult time." And he does not mean because of the refugees, I learn. The difficulties lay in the fact that the Sámi people still did not have stewardship of their own land. They had no say in how pipelines were drawn across their ancient reindeer pastures, who could fish in their traditional waters, what mining companies were allowed to do to the land, and where windmill parks were being built. It had been this way for centuries. Even though there had been some improvements in how the Sámi were treated in the early 21st century, it wasn't until the 2040s that things began to truly change. "And the final change came in 2064, when the IPLCs (indigenous peoples and local communities) took their seat at the UN," says Lihtonen.

→

There are almost 5,000 elders gathered here from all over the globe. The logistics have not been easy, Lihtonen tells me. "It has taken two years of planning." Travel has been done slowly and mindfully of the environment. Food has been brought in the old traditional forms: dried and preserved in different ways.

Could the summit not have been successfully done virtually as well, I ask. Most global summits are, after all, conducted that way.

"It could, and it could not," Laiti says. Lihtonen nods.

"We could have discussed issues and made decisions, sure. But we would have not learnt from each other. We could not have shared stories, and food. We could not have paid thanks together or talked to the land together. And so, the decisions we would have made might have been the wrong ones." Meeting in person is cumbersome for such a huge council, he admits, but he insists on it being absolutely necessary.

The summit started with three days of rites of thanksgiving to the land, to which I was not privy. After that, the daily schedule has been a mix of ritual and meetings about the different stewardship issues faced by the councils around the world. When I ask about this, Laiti smiles. "We do not see it as two separate things. We ask the land for advice on what to do with the land."

Is it really that simple? She looks at me sternly. "Why would it need to be more difficult than that? The land has always taught us how to live. For a long time, people just did not listen."

But now we listen. Since its formation nearly 40 years ago, the GIPS

→

has successfully rewilded huge areas of Europe and North America. More than 70 previously endangered species of wildlife have reached healthy populations. The rainforest restoration programme in South America has already made enormous headway, and dam restoration in Eurasia is, after some initial difficulties, well under way. The projects have been largely funded by the global climate fund. I ask Laiti and Lihtonen what the next big challenge is. They look at each other.

"We have to make ourselves unnecessary," Laiti finally says. "We cannot govern the world for everyone. We must teach everyone to care for it themselves."

"Every human being must re-forge the broken link between themselves and the earth," Lihtonen says. "That is the only way. We are helping during this time of transition, but it is not our job to be nannies forever."

The GIPS has, during this summit, taken the radical decision that its primary goal is to dissolve itself within the next 20 years. "We think the world will be ready by then," Laiti says as she takes me back on the long trek to the shuttle that will take me to the maglev train station, some 120 kilometres away. "It has to be."

I take one last look at Mount Rásttigáisá, the holy mountain which has been teaching and guiding the Sámi people for centuries. And when camp is broken after midsummer, there will not be a single trace of the summit in a month or so, Laiti assures me. "First there will be dead

→

circles where the tents and teepees and yurts and lavvus have stood. But by the end of summer, the grass will once more be green."

Just one more decade to go. **TURN TO PAGE 61.**

We let multinationals buy up all the land in orange zones.

They essentially control the governments and the only state services are funded by them.

- You made it to 2050. TURN TO PAGE 146.

Peace is declared.

After a decade of fighting, society is exhausted. We're ready to do whatever it takes to ensure that war never breaks out again.

We strengthen our commitments to the United Nations. Countries that left it in a huff over the past few decades re-join.

The UN is given new powers. Their first goal is to rebalance the climate, because we recognize that climate change was a major part of inflaming conflict.

The UN launches a number of geoengineering projects, including a fleet of satellites in space that reflect sunlight away.

Society has invested billions into their militaries. We decide to retrain and repurpose them for huge new projects to restore the environment and act as an international disaster response team.

- Over the next few years, things start to look much better for humanity. TURN TO PAGE 262.

The News

All the news, all the time.

3 January 2060

UPDATE YOUR TRACKERS (PAGE 291)

ECONOMY **OH NO...** TEMPERATURE **2.5** CONFLICT **RUN!**

TOP ARTICLE

Anger grows over US sulphur pipe

China leads call for sanctions over unilateral geoengineering

OTHER STORIES TODAY

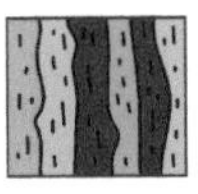

California secedes from the United States as geoengineering causes winter rains to fail; joins Washington and Oregon to form new "eco-utopia" of Cascadia

Eco-terrorist cell from Bolivia kidnaps head of the geoengineering assessment project. President William Trump: "We don't negotiate with terrorists and we will Make the World Great Again"

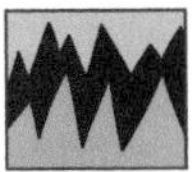

Stocks in sulphur-mining companies see a record increase

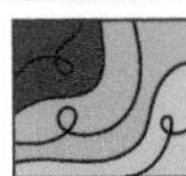

China faces food rationing as vital supply chains from India and South Africa dry up due to drought

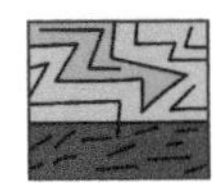

Sahara begins to creep south as the US geoengineering project disrupts global weather

Let's see what the 2060s will bring... **TURN TO PAGE 62.**

The rich have a point. By altering the economy in such fundamental ways, GDP really has dropped in many places.

Which is weird, because most people have seen a marked improvement in their quality of life.

So, obviously how we're measuring the economy is wrong!

The world stops focusing on stuff like GDP and stock-market growth and starts focusing on a new, single GHW (global human welfare) score. This score measures the health, happiness, security and wellbeing of all humans around the world.

With this score, the proposed actions are predicted to dramatically improve the economy.

> *i*
>
> Gross domestic product (GDP) is a tool for measuring the size and health of an economy. But this measure is imperfect, as it does not take into account whether our economies are environmentally sustainable, healthy or equal. Many countries with high GDPs are causing the most damage to our environment. For example, both Saudi Arabia and the United States are within the top ten highest-emitting countries worldwide and suffer from above-average levels of inequality. On paper, these countries would appear to be doing very well due to their high GDPs.

- Congrats! You made it to 2040! TURN TO PAGE 219.

Thanks entirely to a well-argued, impassioned editorial you wrote, rich countries cave to the pressure.

Although they say they can't reduce their own emissions faster, they do contribute trillions of dollars to a global climate fund.

This fund is used to research and deploy technology that will help to stabilize the climate, and to help poorer countries adapt to more extreme weather events.

> *i*
>
> The Green Climate Fund was set up after the Copenhagen climate negotiations in 2009. It is supposed to support the efforts of developing countries in responding to the challenges of climate change. So far, rich countries have failed to meet their promises and given much less money to the fund than they said they would, but there is hope that this might change as citizens put pressure on their leaders to do more about climate change.

- Well, you made it through the decade! TURN TO PAGE 55.

UPDATE YOUR TRACKERS (PAGE 291)
ECONOMY **NO CHANGE** TEMPERATURE **NO CHANGE** CONFLICT **NO CHANGE**

PRESS RELEASE

Vaccine Queen opens the first of 100 tech universities across West Africa

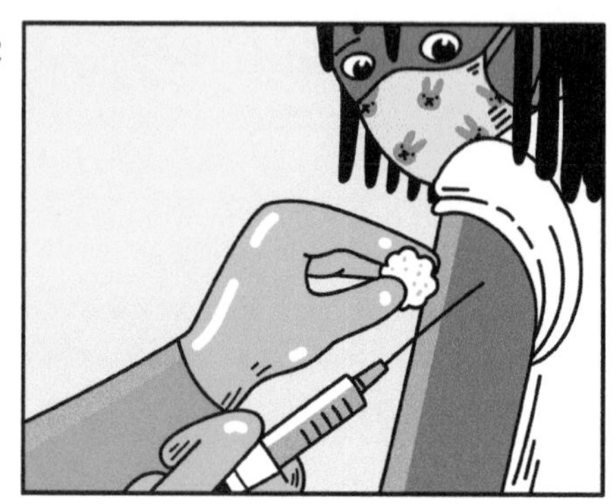

BY LAUREN BEUKES

ACCRA: With the launch of the debut OkC01 University in Ghana's capital today, Nigerian-Australian vaccine magnate, Dylan Okereke-Clifford joins the ranks of billionaire techpreneurs investing in the continent.

OkC01 is the first of a hundred tertiary education institutions with a focus on technology and engineering and a climate-heavy syllabus planned to expand across participating countries including Nigeria, Liberia, Ghana, Ivory Coast, Senegal and Sierra Leone.

Known affectionately as the Vaccine Queen of Lagos, pharmaceutical heir Okereke-Clifford made her fortune post-Covid in manufacturing

→

and distributing vaccines throughout Africa, in partnership with China's HealthEEE.

HealthEEE is proud to continue its partnership with Okereke-Clifford as one of OkC's elite Future Employer partners, who will not only help develop the cutting-edge syllabus, but also have first pick of graduates, guaranteeing 40 per cent employment and ensuring the continent's top minds get the best possible global opportunities they deserve. Other Future Employer partners include the likes of Alibaba, Pfizer and HSBC.

At the ribbon-cutting ceremony in Accra today, Okereke-Clifford said, "I'm thrilled to be able to open new possibilities to West Africans, to expand our minds and our futures. I look forward to seeing OkC graduates taking their rightful places in the global economy and I hope they too will find interesting ways to change the world."

The OkC universities are free to attend and Okereke-Clifford's brand-new ReallyReal™ VR initiative hopes to allow more students to attend virtually from wherever they are on the continent.

OkC01 campus opens in the defunct and neglected Accra Central Library, redesigned and expanded by British starchitect Thomas Heatherwick. "This is another way the private sector is able to step up where government fails," Okereke-Clifford says, adding that the library lives on in virtual reality, and all the books have been digitized. Her company will be distributing ReallyReal™ headsets to all Accra residents to ensure they continue to have free access to all the books.

What will this new decade bring? **TURN TO PAGE 11.**

We drop millions of tons of iron and chemical nutrients into the ocean, hoping to stimulate the growth of phytoplanktons that will trap carbon dioxide from the air and cool the planet.

Unfortunately, this triggers a horrible, unexpected cascade of impacts that no one anticipated. Around 80 per cent of all ocean life dies within two months. WHOOPS!

No one is sure exactly what went wrong, because all the scientists suffocated to death before they could do any studies. So did all other humans on the planet, including you.

- Actually, how are you reading this right now? You're dead! TURN TO PAGE 158.

Large parts of the world are declared "red zones" where humans can no longer live safely.

We talk about "strategic retreat" to safer places, but this creates tensions.

Where should these people try to go?

- Build new migrating cities. TURN TO PAGE 94.
- Retreat to safety. TURN TO PAGE 171.

The News

All the news, all the time.

3 January 2050

UPDATE YOUR TRACKERS (PAGE 291)

ECONOMY **SWEET** TEMPERATURE **2.1** CONFLICT **SNIPPY**

TOP ARTICLE

Millions die as world's first "global syndemic" declared

Deforestation linked to concurrent influenza-paramyxovirus

OTHER STORIES TODAY

Europe's Green First Party achieves majority in EU parliament, immediately passes tough new immigration restrictions

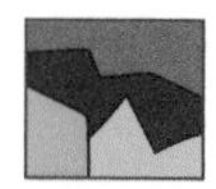

Russia resettles climate refugees in new Siberian camps, refuses photojournalists access

SpaceX offers first long-term leases in space colony

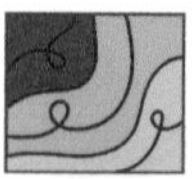

Heatwaves overwhelm crumbling infrastructure, leading to blackouts and deaths

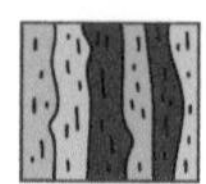

Synthetic coffee, chocolate now cheaper than natural counterparts

Let's see what the 2050s will bring. **TURN TO PAGE 161.**

Strongmen leaders are elected in countries around the world promising to restore traditional gender roles.

They dismantle universal basic income programmes.

In an effort to prove their machismo, a lot of them start small-scale wars and beef up their militaries.

Oh no! You lost your UNIVERSAL BASIC INCOME badge.

- International tensions are heating up. TURN TO PAGE 210.

Corporations argue that they need tax cuts and looser environmental regulations so they can grow, and hire more people.

A series of tax cuts are passed. Corporations celebrate by launching some high-profile flashy social-investment projects, like new schools, hospitals and factories.

But after a couple of years, these promises haven't actually turned into a lot of new jobs.

- We need to try something else. TURN TO PAGE 136.

The News

All the news, all the time.

3 January 2030

UPDATE YOUR TRACKERS (PAGE 291)

ECONOMY **THIS IS NICE** TEMPERATURE **1.4** CONFLICT **SNIPPY**

TOP ARTICLE

Chevron CEO faces life imprisonment for ecocide

Landmark ruling finds executive's actions were premeditated in first test case for new ecocide laws

OTHER STORIES TODAY

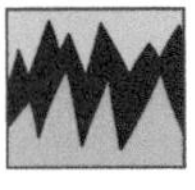

"We tackled Covid, now let's defeat Malaria": Countries agree to keep contributing one per cent of GDP to tackle global health priorities

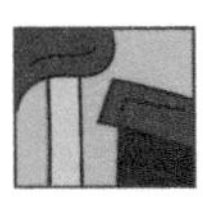

California passes ban on selling and trading animal products: Boom in illicit underground meat restaurants

Read more TURN TO PAGE 196.

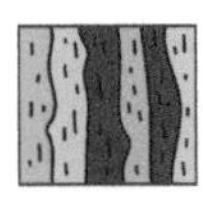

Revitalization of small-town economies and affordable housing: 20-somethings can afford to buy their own homes

First-ever Category-6 cyclone in the southern hemisphere devastates Mozambique

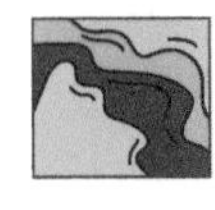

Successful test of Great Barrier Reef corals relocated to Tasman Sea

Global ban on tropical deforestation leads to revival of orangutan population

What will this new decade bring? TURN TO PAGE 249.

We set up a global vaccine-distribution fund that ensures that the most vulnerable are protected from Covid throughout the world.

It's the largest international co-operation project in living memory, and leaves the whole world feeling really warm and fuzzy about humanity's ability to work together.

Covid-19 numbers start to fall, and the people who do get sick tend to be younger, and less likely to be hospitalized or die.

Social-distancing measures start to relax everywhere. It looks like the world will get back to normal within a year.

- Hooray! Time to bust out some non-sweatpants pants. TURN TO PAGE 180.

All of this migration within countries leads to a lot of internal tension between different social groups.

A number of civil wars erupt all around the world.

- A number of strongmen leaders emerge. Most of them are too focused on their domestic conflicts to care about international climate treaties. TURN TO PAGE 57.

The News

All the news, all the time.

24 June 2080

UPDATE YOUR TRACKERS (PAGE 291)

ECONOMY **SWEET** TEMPERATURE **1.5** CONFLICT **PEACEFUL**

TOP ARTICLE

The great sea ice return

Summer fest in the Arctic as Inuit communities celebrate the progression of sea ice returning

OTHER STORIES TODAY

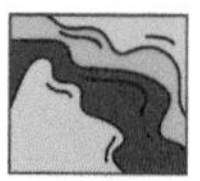

Arctic First Nations school established for intergenerational knowledge exchange about how to thrive in the cold

Report from the Global Indigenous Peoples' Stewardship Summit

Read more TURN TO PAGE 68.

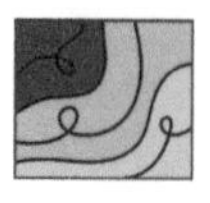

Scientists worried: Did we cool too quickly? Will we need to adapt our adaptations?

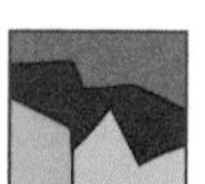

Glaciers have been spotted on Mount Kilimanjaro!

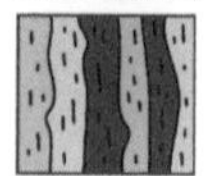

REAL SNOW! Historic ski resort in Argentina opens with real snow for the first time in decades

Just a couple more decades to go. **TURN TO PAGE 61.**

By the late 2020s, economies are battered. Countries need bold new economic-recovery plans. Different groups propose different solutions.

Which economic proposal will your news site give the most time to?

- Governments should invest in large-scale public infrastructure projects, like trains and housing. TURN TO PAGE 18.
- Cut taxes and regulations on businesses so they've got more money to grow. TURN TO PAGE 148.
- Our country first! Let's break with our international trade agreements, put tariffs on imports and stimulate our own economy. TURN TO PAGE 224.

We start experiencing even more climate shocks: hurricanes, fires, droughts, flooding, and animal and insect death.

They're coming fast now.

There's a terrible global economic recession as more and more money is spent responding to disaster after disaster.

- Every country is in crisis mode, focusing on its own problems, far too busy to think about world events. TURN TO PAGE 154.

The News

All the news, all the time.

3 January 2060

UPDATE YOUR TRACKERS (PAGE 291)

ECONOMY **THIS IS NICE** TEMPERATURE **2.3** CONFLICT **SNIPPY**

TOP ARTICLE

"What are they waiting for? People are dying": Humanitarian groups warn of mass famine across South East Asia

Shifting Indian monsoon patterns bring droughts, floods for tenth consecutive year

OTHER STORIES TODAY

Chile holds historic reconciliation summit with climate guerillas: "In retrospect, maybe we should have done more"

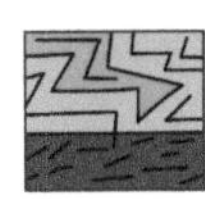

AlphaZero recommends solar geoengineering, world leaders dither

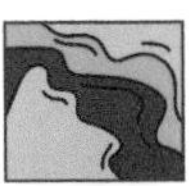

Trans-Arctic shipping surpasses trans-Pacific shipping for first time

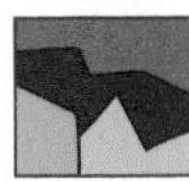

TikTok, Billie Eilish and sea ice: Our favourite nostalgic memories of the 2020s!

Cool, let's see what the 2060s will bring... **TURN TO PAGE 90.**

All this investment in research starts paying off.

We've developed a technology that we think will allow us to engineer the climate and solve global warming once and for all.

There is some risk that something will go wrong. We could spend a few more years doing research to improve the chances that it won't, but we also can't afford to wait much longer.

i

Solar geoengineering is a proposed group of technologies that aim to reflect more sunlight away from the earth to cool the climate. These include ideas like injecting sulphur aerosols into the stratosphere, brightening the clouds above the ocean, or even deploying space mirrors. None of these technologies are ready for deployment (in 2021). Research suggests that solar geoengineering could reduce temperatures, but it could also introduce some big new risks, like disrupting rainfall in major food-producing regions. Solar geoengineering has been called a stopgap measure because it does not reduce greenhouse gas emissions, the underlying driver of climate change. It treats the symptoms rather than curing the underlying condition. Solar geoengineering is highly controversial, with some people arguing more research should be done on these technologies, and others saying that tampering even more with our climate system is too risky.

→

Should we try a solar geoengineering solution now, or do more research?

- Let's keep talking about it for a few more years. TURN TO PAGE 104.
- Let's give it a go. TURN TO PAGE 236.

Concerned about the lack of space in the cities that have managed to adapt to climate change, migrants increasingly build new migrating cities: floating artificial islands that can migrate with the weather.

There are also experiments building new cities in previously hostile environments, like the Arctic, and underground.

- You survived the decade! TURN TO PAGE 63.

UPDATE YOUR TRACKERS (PAGE 291)

ECONOMY **SWEET** TEMPERATURE **1.5** CONFLICT **SNIPPY**

TOP ARTICLE

Gates: Covid-19 approaching "final mile" of elimination

Experts question continuing neglect of other global health priorities

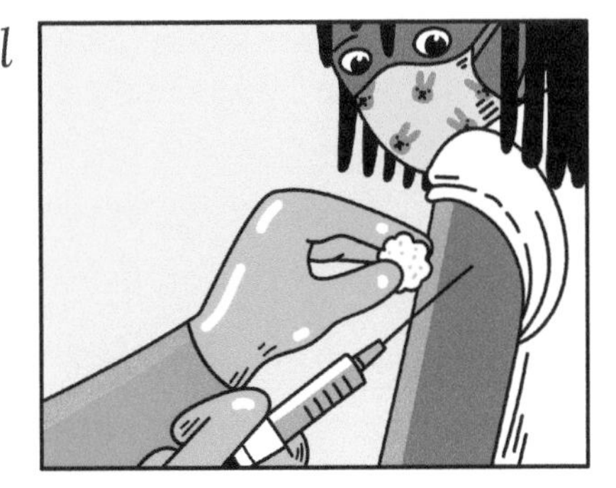

OTHER STORIES TODAY

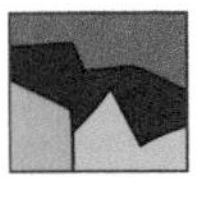

"Economy First": US President Mike Pence pledges to bring prosperity to Appalachian shanty towns by revitalizing the last coal mines

Top Bollywood stars establish fund to reunite families separated by 2027–29 quarantine

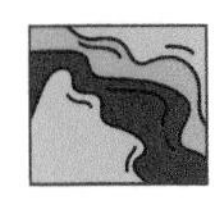

"The reefs are f-cked" admits leading coral reef expert

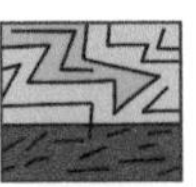

Record-breaking Hurricane Omega devastates Ireland

Vampire bats seen in New Orleans: Experts say no risk, but possible benefits if they eat dengue mosquitoes

What will this new decade bring? **TURN TO PAGE 113.**

As people get wealthier, they use a LOT more energy.

So, even though they're switching to renewable sources, they're using much more energy overall.

This leads to a much fairer society, but delays the transition to renewable energy sources.

Award yourself the UNIVERSAL BASIC INCOME badge (page 289)!

- You made it through the decade! TURN TO PAGE 55.

Cool!

Award yourself the KUMBAYA badge (page 289).

I'm sure this decision won't be important later.

- You made it to 2070! TURN TO PAGE 247.

At the climate summit, leaders need to agree to stricter emission-reduction targets.

There's a debate about whether historical emissions should be counted towards a country's total target (which would damage the economies of currently rich countries) or whether only current emissions matter (which would impact China and other emerging economies).

> *i* Although the whole world suffers because of climate change, some countries are far more responsible for it than others. The US alone has caused 25 per cent of historical emissions. Some believe that historically high-emitting countries should be entitled to a smaller portion of the remaining carbon budget and should compensate low-emitting countries for the damage they've caused.

How hard should we push rich countries to pay for climate change?

- We can't push too hard, or we risk rich countries just pulling out of the process completely. TURN TO PAGE 108.
- This historical emissions proposal sounds ludicrous! TURN TO PAGE 127.
- Let's push! We have no more time to waste. TURN TO PAGE 77.

The News

All the news, all the time.

3 January 20

UPDATE YOUR TRACKERS (PAGE 291)

ECONOMY **SWEET** TEMPERATURE **1.8** CONFLICT **SNIPPY**

TOP ARTICLE

Agribusiness launches drought-resistant millet seed, saves world from hunger as wheat yields collapse

Partners with Amazon-Chase to extend credit to farmers who can't afford to buy seeds outright

OTHER STORIES TODAY

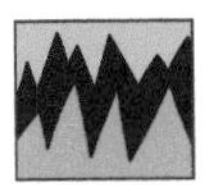

Antitrust court rules Amazon's purchase of JPMorgan Chase is "totes fine, what could possibly be bad about this?"

Natural-gas extraction leads to extensive groundwater contamination in rural areas as regulators defer to energy interests

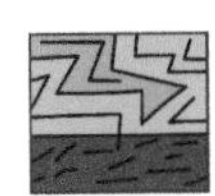

Another hurricane devastates American east coast; OANN host goes on 20-minute live-air rant blaming Mexico for climate change, says they should pay for damages

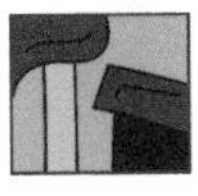

The promise of a good meal: Musk promises room and board in exchange for menial labour in his southern California TeslaTown

The safest escape: How one VR programme can improve your mental health

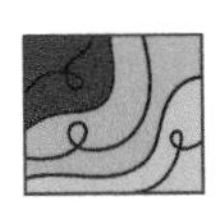

Second dust bowl created in Midwest US as Ogallala aquifer depleted

Everybody's feeling optimistic about the 2040s! **TURN TO PAGE 227.**

Wow, uh, that was lucky.

Six months later, the scientists walk out of their lab saying they've genetically engineered a type of plankton that absorbs and destroys greenhouse gases. The plankton is very cheap to produce, you can perfectly control its rate of reproduction, and there are zero negative side effects.

Over the next few years, we deploy the plankton into our oceans. Temperatures return to where they were before the industrial era, extreme weather events become much less common, and the polar bears have food again.

Everybody is so happy that we've solved climate change, the world's leaders declare a three-day holiday to celebrate. A huge statue is erected in your honour. At the unveiling, your high school English teacher gets on stage and tearfully says, "I was so wrong when I said you'd amount to nothing!" Just then, she pulls off her face and you realize she's a giant plankton, and suddenly your ex is there, who asks you if you want to go ice-skating...

- BRRR! BRRR! BRRR! TURN TO PAGE 135.

Cities welcome new migrants and invest in public housing, built with green technologies and incorporating a lot of urban gardens and social spaces.

Cities become denser, more multicultural. This leads to a new reckoning in what we value, and a greater awareness of our responsibilities to each other.

More people want to live in these cities, where the quality of life is better. Food production is increasingly densified, leaving more rural land available for use.

What should we do with all of this former agricultural land?

- The state should buy it back and rewild it. TURN TO PAGE 54.
- Allow the land to remain in the hands of private individuals. TURN TO PAGE 267.

A series of wars are started over who gets to control the new orange zone territories.

While they're hard to live in, they might be of strategic importance down the line!

A series of small-scale wars erupt around the world that are really proxy wars about which of the world's superpowers have more power.

- You made it to 2050. TURN TO PAGE 40.

The News

All the news, all the time.

12 January 2080

UPDATE YOUR TRACKERS (PAGE 291)

ECONOMY **THIS IS NICE** TEMPERATURE **3** CONFLICT **PEACEFUL**

TOP ARTICLE

The Council is still the greatest!

Successfully averts near-disaster, saves humanity again

OTHER STORIES TODAY

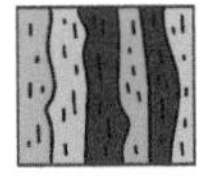

New breakthrough in air-to-food technology will mean we never have to rely on unreliable sunlight to grow our food, ever again

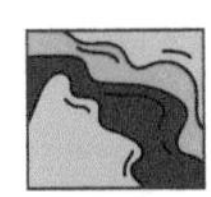

Too hot where you live? Try one of these designer cool suits that will let you go outside for up to six hours

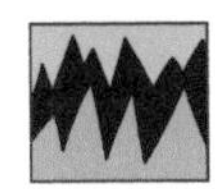

Trust is stronger than blood: An interview with Babylondon mayor Lena Lenin

Read more **TURN TO PAGE 211.**

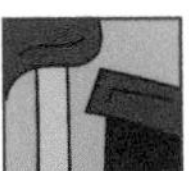

A hundred of the best new underground nightclubs! (literally underground)

Five steps to create an indoor tropical paradise

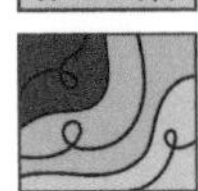

Council reveals plans to launch huge new solar disc into orbit, increasing solar energy supplies hundredfold

Just a couple more decades to go. **TURN TO PAGE 39.**

We choose to do more research, and keep debating, and talking, and talking, and talking, and stalling.

By the end of the decade, we haven't done anything yet.

- What?! It's 2060 ALREADY?? TURN TO PAGE 91.

Only a handful of countries welcome international climate refugees, most of whom are corralled into permanent refugee camps.

It's a humanitarian disaster.

Leaders in rich countries grow increasingly nationalistic, arguing that refugees must be kept out at all costs. They warn that refugees will bring disease and crime into their countries, and that it's the problems of the refugees' home countries that created this whole mess to begin with.

> *i* Eco-fascism is an ideology that blames environmental breakdown on marginalized communities, arguing that overpopulation, over-industrialization, and immigration are causing climate change. Followers often believe that racial purity and anti-multiculturalism are solutions to the climate crisis. Many eco-fascists can be characterized as white supremacists.

- Your correspondents from poorer countries report that there's a lot of anger building. TURN TO PAGE 20.

UPDATE YOUR TRACKERS (PAGE 291)

ECONOMY **NO CHANGE** TEMPERATURE **NO CHANGE** CONFLICT **NO CHANGE**

PRESS RELEASE

Finnish children celebrate first snow day in 50 years

BY MARIA TURTSCHANINOFF

Children in the southern part of the territory formerly known as Finland had the day off from school yesterday, as there was snow for the first time in 50 years.

"I wish my parents could have seen this," said Juhani Larson from Helsinki. I reach him virtually at the school, which is now open again. "My father used to talk about how his grandmother taught him to ski in the 2020s. There hasn't been snow this far south since then."

Snow disappeared from Finland entirely in the late 2040s. It made its return to the northern parts in the last decade, but this is the first time in 50 years that southern Finland has seen snow. It did not snow much,

→

and it is expected to melt in the next few days. But it is a very promising sign.

"I have been teaching the children the history of their country. We have drawn skis, talked about what snow meant for brightening the dark winter months, how the animals used to change into white winter coats. They have studied the drastic increase in pests on crops and in forests alike as the snow and cold disappeared. We have watched old movies with snow and visited pre-warm Finland virtually. But nothing beats feeling the flakes fall on your face and catching them on your tongue. I am so happy I got to experience this myself." He grows misty-eyed as he looks out the window of the log cabin where the school is held, at the children playing outside, building snow forts and throwing snowballs. "One of my favourite comics growing up was *Calvin and Hobbes*. I always wanted to build a snow fort and go sledding. I wasn't sure I would ever see the day when that was possible, right in my home town. And now these children will grow up with snow as part of their everyday winters, once again." He glances out the window again. "I'm sorry, can we wrap this up? There's a snowball fight going on, and I sure would like to join."

You made it to the end of the century! **TURN TO PAGE 185.**

Rich countries threaten to drop out of the summit completely if pushed too hard, arguing that they're still recovering from the post-Covid economic crash.

We decide it's more important to keep everyone in the process than to pass aggressive targets now. We decide to pass laughably small targets that we aim to increase in future.

- Well, you made it through the decade! TURN TO PAGE 166.

So, where do we end up by the end of the century?

We avoided some of the worst effects of climate change, but we completely lost any hope for democracy or fairness.

Unfettered disaster capitalism and an over-reliance on technological fixes created a blindingly unequal world, where the rich are mostly insulated from climate impacts and there's a huge, suffering underclass.

Better hope they never rebel.

How did you get here?
You allowed capitalists to run rampant and did not ever address inequality.

To take action **TURN TO PAGE 268.**

Play with the data behind your decisions at tinyurl.com/2p89vdnu

Unfortunately, the natural methods aren't enough.

We damaged the climate so much in the past that we need more than rewilding to stop these extreme weather events from happening.

> *i*
> Planting trees to absorb carbon dioxide is often celebrated as a win-win climate solution. But scientists warn that planting trees alone will not solve climate change. In fact, there are potential hidden dangers in mass tree-planting schemes: if the wrong trees are planted, they may not absorb that much carbon and could have negative impacts on biodiversity, landscapes and livelihoods. Focusing on planting trees for climate benefits may lead to competition with other land uses, like farming for food.

- We'll need some miracle to stop people from dying. TURN TO PAGE 173.

The News

All the news, all the time.

12 July 2080

UPDATE YOUR TRACKERS (PAGE 291)

ECONOMY **DISASTER** TEMPERATURE **4** CONFLICT **DANGER!**

TOP ARTICLE

Violence erupts in poverty belts around New York and San Francisco

Millions are left without refuge during the peak heat of the summer

OTHER STORIES TODAY

Chinese colonies in Africa are without water after American navy attacks Antarctic ice cargo ships

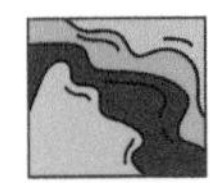

Chinese navy occupies New Zealand waters after fisheries collapse in the tropical Pacific

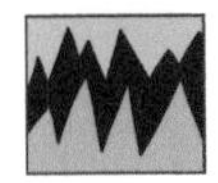

Maize crops fail for the eighth consecutive year, five billion people rely on international food assistance

Millions of climate refugees are offered asylum in exchange for work in Antarctic uranium mines

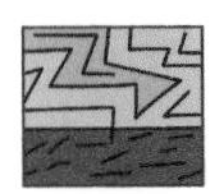

Serious destabilization of Arctic methane clathrates triggers tsunamis in northern seas

Displaced urban populations finally learning sustainable living among indigenous peoples' forest refuge

Just a couple more decades to go. **TURN TO PAGE 30.**

After decades of defunding, democracies have no ability to govern anymore.

Capitalists aren't perfect but we trust them more than big government. We decide that private companies should keep running these projects.

Runaway effects of the climate hacks devastate parts of the world, forcing the last remaining free societies into corporate cities, where they hope to "earn" citizenship through work... maybe. One day. The unluckiest are forced into jobs on the nascent Mars colony.

- You made it to 2080! TURN TO PAGE 235.

There's a resurgence of the climate rebellion movement.

Grassroots groups campaign for change. Millions join a global day of protest, once again led by schoolchildren.

The movement is made up of disparate voices calling for different solutions.

Which of these solutions will your news site focus on?

- We need our countries' leaders to act. TURN TO PAGE 26.
- We need more passionate climate activists getting involved in local politics. TURN TO PAGE 245.
- Business leaders should drive change. TURN TO PAGE 84.
- You heard a rumour that these "protests" are being funded by our national enemies and are just an attempt to undermine democracy. TURN TO PAGE 233.
- Hooligans! These protestors are just looking for an excuse to riot. Let's publish photos of isolated incidents of vandalism and not mention their proposed solutions at all. TURN TO PAGE 238.

As women become better educated, birth rates plunge all around the world.

Economists panic at the looming "old age" bulge.

How should we respond to this economic issue?

- Invest in social care. TURN TO PAGE 237.
- Let's ban birth control. TURN TO PAGE 157.

UPDATE YOUR TRACKERS (PAGE 291)
ECONOMY **NO CHANGE** TEMPERATURE **NO CHANGE** CONFLICT **NO CHANGE**

PRESS RELEASE

US-sponsored non-state actors seeding oceans with banned geoengineering tech?

(Leaked transcript of video call between US Leader-for-Life and African President of the Free States)

BY RAJAT CHAUDHURI

US President, **William Trump** (A dim tantalum bulb swings slowly above his head, ominously close. Shadows dance on the bare wall. The light goes off repeatedly and every time this happens, a secret-service guy rushes in with a torch.)

African leader, **Virendra Xulu** (Dazzling backdrop of a solar farm stretching right up to the horizon. He seems to be floating on an infinity pool.)

TRUMP

Hello, Chef! How you doin'? Hope you don't mind me calling you Chef, Mr President of the Free States?

XULU

Howzit, Bill? Can't see you clearly.

TRUMP

That's amazing! Know what, we had a little setback but are getting back into the electricity game. It will take a little more time and some help from friends while we sweep the States clear of hostile forces.

XULU

Good to hear that.

TRUMP

Great, thanks, Chef. But coming back to business, you've done a classy job with the solars I'm told. And hey, I can see you've got a lot of water over there too.

XULU

Zilch! This pool is just holo-AR, but the solars, yes. Those are real... I'm told.

→

TRUMP

Secretary of Defence Maven tells me he has already informed you that we need that electricity you are pumping out from the sun. For our bioreactors, while agri gets back on its feet. We are going after the losers who engineered that blight.

XULU

Happy to help, but we will be left with no power if—

TRUMP

—You have the crop fields and the elephants. Why do you need power? We can use it here for our all-American ThinAir process. My scientific adviser, Clara, tells me we got loads of the dioxide here... same carbon-fucking-dioxide... and the micro-bees; we only need your sun volts to manufacture our food.

XULU

Whoa! The bees? Hmm... but Bill, that tube you stuck into the Milky Way – that was what started all this trouble and the wars. My northern regions turned wasteland and the Sahara is still crawling southward...

TRUMP

Let it crawl. It's all China, so no harm done. It was the commies that started this business. Besides, you have enough sun over there; a little darkness at night is good rest for your eyes.

→

XULU

But the sulphur you are pumping up—

TRUMP

—Look here, Chef. You are a fantastic guy, I admire your leadership. But the Great Gustaff [SO_2 pipe] is non-negotiable. We can do some other deal... lemme see... I hear you don't see any fish lately?

XULU

Ag, man, not a shoal in our oceans.

TRUMP

Great. We can bring those back.

XULU

I always knew you were a witch-doctor, Bill, but if you take down the Gustaff—

TRUMP

—We will vaporize those Mexican terror groups seeding the oceans and driving away the fish and you give us electricity.

XULU

Mexicans? Please don't nuke 'em again...

→

TRUMP

So it's a done deal. Our floating Tesla towers will get your sun volts across the pond and you enjoy grilled snapper, like good ol' days.

XULU

But Bill, Mr President, your people will get respite from the acid showers if you bring down that tube of yours!

TRUMP

I will tweet that out, professor. Meanwhile enjoy the sun while you have it!

(Drone attack alert can be heard in the background.)

Onwards into the 2070s... **TURN TO PAGE 29.**

The Council scrambles to develop pollination drones and heat-resistant crops, but everything happens too quickly.

There's not enough food. Our glorious civilization falls apart and most people starve. Some remnants of human society survive, but unfortunately, you do not. A cushy life of being the editor of a news site did not prepare you well for survival in a hostile post-apocalyptic wasteland.

- Actually, how are you reading this right now? You're dead! TURN TO PAGE 158.

The Global Climate Council argues that they need stronger enforcement powers to protect the world, and effectively becomes the world's (highly authoritarian) central government.

The leaders of the rogue states that attacked the space fleet are imprisoned and the Council takes control.

The Council appoints a Media Fairness Monitor to your newsroom. She won't tell you her name, and she never smiles. She will review everything you publish from now on.

Award yourself the COUNCIL LOVES ME badge (page 289).

- You made it to 2070! TURN TO PAGE 262.

UPDATE YOUR TRACKERS (PAGE 291)

ECONOMY **NO CHANGE** TEMPERATURE **NO CHANGE** CONFLICT **NO CHANGE**

PRESS RELEASE

Exile nation: What happens when you're forced out of paradise?

BY LAUREN BEUKES

In the dark of the Qatar desert, Dr Shanaaz Naicker sits on the edge of the rickety corrugated rooftop of a former warehouse and looks out at the glowing beacon of Bezosistan, that Amazon paradise zone 50 miles away she used to call home until her unexpected retrenchment nine days ago. "I used to have a whole damn life there," the 27-year-old bioengineer sighs. A two-bedroom apartment in the sky garden district with an enclosed balcony for her cats, Hamilton and Angelica. Her work was important and useful, developing organ-growth tech, although she can't get into the details because of the "hectic" NDA she had to sign. She used to kayak to work along the canals, she had

→

interesting and accomplished friends, was a regular at the local vegan deli, and she'd recently started dating an AI project manager, Emad, who declined to be interviewed for this article.

Now Shanaaz is one of the "warehoused" of Limbo City – the unemployed (and maybe unemployable), camping out among the abandoned warehouses on the edge of Bezosistan's borders, sending off their resumes and hoping, fervently, to get back in.

It's not the worst place you could live. Sure, the warehouse is crumbling, subject to the ravages of the desert, and seasonal infestations of camel spiders, but the community is made up of some of the brightest young rejects who have used their considerable tech skills to jerry-rig a pale simulacrum of the lives they used to have inside the paradise zone.

Shanaaz walks me through the highlights. There's an allotment greenhouse, and a meat growlab that used to run on solar power until Qatar's sulphur phosphate injection into the atmosphere dimmed the sky, so now the spin bikes and treadmills in the makeshift gym are hooked into the DIY power grid. There's a shared kitchen where people cook together, a woefully under-resourced robotics lab, a small clinic, and even a sculpture park with DIY creations – welded ironwork, wind-harvesting mechanical beasts – starting to colonize the stretch of desert out the back where Shanaaz has set up her tent. She's elected to have some privacy rather than take a bed in one of the dorm rooms. Most importantly, there is wifi, so the Limboites can send out job

→

applications, endless job applications, almost always to Amazon and their affiliate corporate states.

The reason they don't look further afield is that very often they can't. Many former Amazon employees, like Shanaaz, are bound by brutal non-disclosure agreements and restraints of trade that make it almost impossible for them to emigrate to another corporate state, or at least not with their current skill set. It's why she's signed up for a coding course offered by one of the other residents, and why she's, in turn, offering bioengineering 101 to her commune mates, 9 a.m. to noon weekdays. The rest of the time she works in the clinic, hoping that practical hands-on medicine will be an asset, another tick box on her resume.

We wend our way through the outdoor bazaar consisting of small local businesses, stalls selling electronics parts, and food trucks set up in shipping containers at the edge of the sculpture park. A young man in a dark suit is playing guitar on a raised stage strung with fairy lights and singing sultry lounge covers of hits through the decades. "Smells Like Teen Spirit" segues into "Old Town Road" and an acoustic version of "WAP", occasionally interrupted by the low buzz of drones flying in, hung with packages.

We get coffee from Outer Perimeter, one of three vendors here, and the only one without trellis tables populated by people with laptops typing up cover letters and resumes or appeals to HR.

"Jamala likes people to grab their coffee and go," Shanaaz explains.

"It's takeaway," the tall Congolese man says, "so you must take it away!" Jamala is a former VR programmer, who has been in exile from Bezosistan for 19 months already. He specialized in programming realistic grass that moves with the wind, "and now I make realistic coffee," he grins. "Spoiler. It's not coffee, it's chicory."

"Everyone's got a theory about why Bezosistan doesn't come and shut us down," Shanaaz says. "Maybe they feel guilty for retrenching us. Or they're keeping us on standby for when someone dies or screws up, and they want to keep us within easy reach if they want to rehire us. It doesn't happen often, but there was this one guy."

"Ari Lacrosse," Jamala chimes in.

A local legend, the man who escaped, who reclaimed the dream. But the details vary depending on who you talk to: when he was here, the job he returned to, how he did it. It was three years ago or four. There are darker rumours: that he was a programmer who sabotaged his own code so ingeniously he was the only one who could debug it and they had to hire him back. Or he was a lawyer with dirt on a competitor, or an HR manager with a blackmail catalogue of buried sexual harassment cases, a physicist whose replacement mysteriously died of food poisoning. But Shanaaz waves those theories away as "melodrama and revenge fantasies". She takes the lid off her coffee to stir, and I smell the whisky coming off it. A friendly cat swirls between her ankles. "He was good at what he did and they needed him back. You're irreplaceable, until you're not."

→

"They can't shut us down, in case they need us," Jamala agrees.

But perhaps there's another reason Limbo City is tolerated, I think as I notice another drone buzzing in overhead, the Cheshire smile of Amazon's logo on the box. After all, they're customers too.

It's almost time for me to leave. I bend down to stroke the cat, scruffy in the way of the very recently neglected, like so many of the inhabitants here. "Is this one of yours?" I ask. The cat nudges its head into my hand, purring.

"No. I had to leave them behind," Shanaaz says, sipping her spiked coffee and staring out at the bright glow on the horizon. "But I'll be back for them. Soon. You'll see."

Let's see what the 2050s will bring. **TURN TO PAGE 161.**

The EU and America claim that the "historical emissions" proposal is nothing but a power grab by China.

Global tensions are inflamed further. Everyone walks away from the summit angry and no resolution is passed.

- Well, you made it through the decade! TURN TO PAGE 199.

We drop a couple of nukes; why not?

But things are already so awful it doesn't really change that much. On the day the largest bomb is dropped, more people die of the infectious diseases that now run rampant.

- Well, you survived the decade. TURN TO PAGE 207.

Rich countries argue this is unfair.

There are far too many migrants. Their entire way of life will be disrupted if they have to do this.

- A number of strongmen leaders emerge in rich countries that say they'll pull out of the international migrant treaty. TURN TO PAGE 57.

UPDATE YOUR TRACKERS (PAGE 291)

ECONOMY **NO CHANGE** TEMPERATURE **NO CHANGE** CONFLICT **NO CHANGE**

PRESS RELEASE

Reddit thread: Globohomo vaccine whistle blown

Online Culture reporter Sophia Al-Maria brings us a sampling of what's happening on your favourite social networks.

BY SOPHIA AL-MARIA

r/conspiracy · Posted by u/curveddelirium 1 hour ago

GLOBOHOMO VACCINE WHISTLE BLOWN!

FOR ALL THOSE WHO WANT TO KNOW THE TRUTH

YOU NEED TO READ THIS

Liberals have been trying to push their agendas like LGBTQ "rights" etc. down our throats via the liberal fake news media. Forced diversity and Covid are the #1 news stories. I have NEVER worn a mask and I am 100 per cent healthy and heterosexual.

Given how feminism has demonized males and masculinity and tried to make all men into beta cucks. Now they want to literally VASECTOMIZE and HYSTERECTOMIZE heterosexuals covertly via the vaccine so they can reduce the population size and make it easy to govern us under one unified government aka the NEW WORLD ORDER. The population is WEAK and DISTRACTED.

Now as to the vaccine... think about it. It's the perfect delivery service. It will be rolled out to BILLIONS of people GLOBALLY. The media industrial complex has filled everyone with fear that if you walk outside without a mask you're going to kill 100 grannies. This is going to be ENFORCED under the guise of "protecting the world from deadly Coronavirus" when in reality the deep state wants to kill two birds with one stone. The vaccine will cause all who take it to have a SERIOUS and IRREVERSIBLE aversion to heterosexual lifestyles to STOP BREEDING and CONTROL the population.

I have seen some theories floating around on here that the "virus" was lab-designed and strategically released. Even if that were true you would have to be dumb to ignore the fact that it came from China AKA the same country that imposed a one-child policy on their citizens to CURB POPULATION GROWTH. Coincidence? Of course NOT! They are trying to STERILIZE US!!!

WAKE UP!!!

→

circularimmunity363

> this makes total sense. we all know that the clinton crime family are involved in satanic rituals and sex slavery. the majority of this is homosexual

defeatedloathing

> GET WOKE GO BROKE

firmfetish83

> The deep state is evil and Satanic. LOOK IT UP IF YOU DON'T KNOW

ManicalIInfiltrator

> the good news is all u incels don't have to worry bc no one wants to have sex with u losers anyway LOOOOOOL

fitkna

> Don't trust this guy, he's a shitlord. I've seen him on other threads trying to spread this trollfuckery

shoulder_to_the_wheel

> I think Gallium nitride (GaN) is the medicine for covid-19

oopslord9

> Wiki: "Bulk GaN is non-toxic and biocompatible.[49] Therefore, it

→

> may be used in the electrodes and electronics of implants in living organisms."
> perfect for implanting chips in humans???

adzmk3

> i want to believe but 2 c more proof does any1 have linkz?

HarmlessNightmare

> yes check out my youtube page i explain everything i have many videos talking about how the deep state already controls us watch my videos but get ready once you know u can't unknow lol

PM_ME_UR-NOODZ

> the soy boys are gonna luv this

crylikeababy

> which one of u incels is gonna be chad enough to resist?

dynamic_oracle

> Even Reddit is controlled by China

ConcreteDefection

> "The best way to control the opposition is to lead it ourselves."
> — Lenin

What will this new decade bring? **TURN TO PAGE 11.**

Cool!

I'm sure this decision won't be important later.

- You made it to 2070! TURN TO PAGE 247.

Yeah sorry that was a dream.

- You check your phone and see you've got three text messages from your secretary reminding you that they still need a decision on that editorial. TURN TO PAGE 20.

Fossil energy companies blame the aggressive climate emission laws for high levels of unemployment.

They say that they can create a lot of new jobs in fracking, and that can be a transitional energy source while we move off coal. They can even re-invest those profits into carbon capture technology! A little more pollution now, but it will give us much more money to spend fixing the problem in future.

> *i* Carbon capture and storage technologies aim to remove carbon dioxide pollution from the air and store it where it can't re-enter the atmosphere, usually underground. Many of these technologies are still energy-intensive and expensive, and none are yet able to work at the large scales that would be needed to stabilize the climate.

Should we approve new large-scale fracking projects?

- No. We can't afford to pollute the air even more. TURN TO PAGE 230.
- Yes, but only if the energy companies invest some of their profits in capturing the pollution that's created. TURN TO PAGE 138.
- Yes, but let's nationalize this and use the profits to fund a universal basic income. TURN TO PAGE 96.

The News
All the news, all the time.

9 August 2090

UPDATE YOUR TRACKERS (PAGE 291)

ECONOMY **ROCKING!** TEMPERATURE **1.5** CONFLICT **PEACEFUL**

TOP ARTICLE

"We did not let hopelessness paralyse us," says Xhosa elder in passionate speech reflecting on environmental restoration

Celebration as global temperatures return to 2020 levels

OTHER STORIES TODAY

Solar-powered Cape-to-Cairo magnetic levitation train "derails" mid-journey in the Congo basin, causing severe impacts to the home of mountain gorillas and creating mass outcry across the world

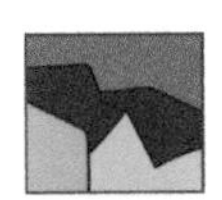

Finnish children celebrate first snow day in 50 years
Read more TURN TO PAGE 106.

DNA testing: Variety of ancestral roots gives us a new sense of place as planetary citizens living in our local environments

Species mass-migrate back to their original biomes as environmental restoration and climate stabilization has enabled viable habitats

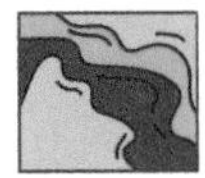

Penguins celebrate as polar bears are repatriated to the Arctic; bear custodians say "Yes yes, we're putting them on a diet"

You made it to the end of the century! **TURN TO PAGE 185.**

Energy companies have to use a portion of their fracking profits to research new ways to reduce climate change, or help the world adapt to it.

This sparks a lot of innovation, and mints a whole new generation of tech billionaires and green influencers. Green is the new gold, baby!

- Congrats! You made it to 2040. TURN TO PAGE 99.

Climate rebels around the world start implementing small-scale projects to hack the climate.

These projects somewhat work, but they are being used at a time when there is already massive disruption to weather systems. Thanks to climate change, communities have had to adapt to a lot of sudden weather changes: areas that had a lot of rainfall suddenly have to cope with droughts; cities that had become more resilient against droughts suddenly have to deal with flooding. Now, with climate-hacking projects, nobody is sure just who or what is responsible when something bad happens.

Overall, the temperature of the planet doesn't increase as much as it might have, but the disruption to people's lives is arguably higher.

- You made it to 2080! TURN TO PAGE 239.

The News

All the news, all the time.

9 August 2090

UPDATE YOUR TRACKERS (PAGE 291)

ECONOMY **OH NO...** TEMPERATURE **3.7** CONFLICT **DANGER!**

TOP ARTICLE

New conflict erupts on California–Mexico border

Communities battle over control of few remaining rain catchment areas

OTHER STORIES TODAY

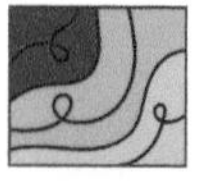

Lightning strikes twice! Psalm West wins another child permit in parenting lottery (lucky bastard)

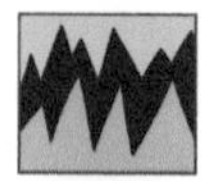

Latest poll: Thirty per cent of global population hopes to move to Mars

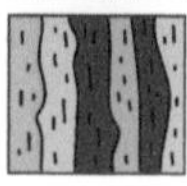

"I was a child soldier": How one teenager survived the Patagonian war

Thousands complain as heatwaves cripple mobile-phone use outdoors, teens mock Gen Z for not using refrigerated phone cases

Bats drive pigeons out of Kingston, Jamaica, take over kudzu-choked remains of the sunken city

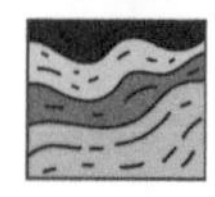

Last holdouts finally abandon the Sahel

You made it to the end of the century! **TURN TO PAGE 172.**

Uh-oh.

Remember all the billionaires who left in the 2060s and went to live on a Mars colony?

They just invaded Earth. They say they want their planet back.

Oh boy. Do you have that KUMBAYA *badge?*

- Ah. Yes. We disbanded all of the militaries, remember!? TURN TO PAGE 147.
- Nope. We kept some militaries in place, so we can fight them off. Phew. TURN TO PAGE 137.

Phew, is it just me, or is it getting pretty hot in here?

Despite all we've done, the world is really starting to feel the effects of climate change. 2041 is a particularly brutal year: so many disasters happen all at once that many of them don't even make it onto the front page of your news site.

The disasters displace millions of people. In the US, droughts and hurricanes force people north. Many of them lose their homes and livelihoods. Bangladesh has become largely uninhabitable due to constant flooding. Failing fish stocks in the tropics lead to near-starvation among communities that depend on fishing to survive, and force people into cities. Plans have been made for large parts of Hong Kong, Miami, Lagos and Manhattan to be abandoned to the rising sea water.

R.E.M.'s classic song "It's the End of the World as We Know It (and I Feel Fine)" is the most listened-to song of the year.

Millions want to resettle in more protected places. Rich countries recognize that they have ageing populations and new immigrants will be a huge boost for their economies, but resettling people will worsen housing shortages in the most-desirable cities.

i

As climate change leads to rising sea levels and extreme events like repeated heatwaves, droughts and floods, many parts of the world are becoming harder

→

to live in, forcing people to leave their homes. Over a billion people live in areas exposed to land degradation or climate-related sea-level rise. Most of this migration will happen within countries, not between countries.

How should society respond?

- Let's build new public housing in cities. TURN TO PAGE 101.
- Let's offer tax breaks and loosen regulations so that private businesses build more housing. TURN TO PAGE 263.

Phew, is it just me, or is it getting pretty hot in here?

The world is really starting to feel the effects of climate change. 2041 is a particularly brutal year: so many disasters happen all at once that many of them don't even make it onto the front page of your news site.

The disasters displace millions of people. In the US, droughts and hurricanes force people north. Many of them lose their homes and livelihoods. Bangladesh has become largely uninhabitable. Failing fish stocks in the tropics lead to near-starvation among communities that depend on fishing to survive, and force people into cities. Plans have been made for large parts of Hong Kong, Miami, Lagos and Manhattan to be abandoned to the rising sea water.

R.E.M.'s classic song "It's the End of the World as We Know It (and I Feel Fine)" is the most listened-to song of the year.

Millions want to resettle in more protected places. Most resettle in their own country, but many also cross borders in an attempt to give their families a better life.

i

As climate change leads to rising sea levels and extreme events like repeated heatwaves, droughts and floods, many parts of the world are becoming harder to live in, forcing people to leave their homes. Over a billion people live in areas exposed to land degradation or climate-related sea-level rise. Most of this migration will happen within countries, not between countries.

How should society respond?

- There are just too many people. Let's build refugee camps. TURN TO PAGE 105.
- Every country should have to accept a number of refugees proportional to their carbon emissions. TURN TO PAGE 129.
- Let's make it easier for businesses to sponsor working visas, so people can only move to a country if they have a job there. TURN TO PAGE 234.

The News

All the news, all the time.

3 January 2050

UPDATE YOUR TRACKERS (PAGE 291)

ECONOMY **THIS IS NICE** TEMPERATURE **2** CONFLICT **TENSE**

TOP ARTICLE

Last chance to see: Five cities expected to sink within the next decade

Booming disaster-tourism industry gives morbid travellers the last chance to experience cities threatened with sea-level rise

OTHER STORIES TODAY

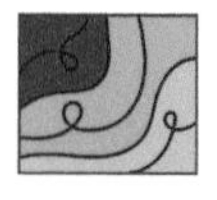

Montreal Protocol banning ozone gases dissolved as US pulls out; swing vote Amy Coney Barrett rules that treaty is an infringement on individual liberties of corporate persons and sovereign states

Exile nation: What happens when you're forced out of paradise?

Read more TURN TO PAGE 122.

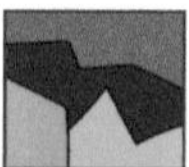

Missing the cold? New frozen theme parks bring a welcome chill to your bones

The sounds of summer: How one project is archiving the changes to what we are – and are not – hearing when we walk outside

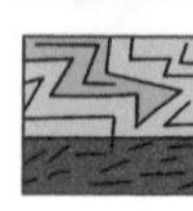

85-year-old Jack Ma proposes bold plan to re-freeze the poles

Let's see what the 2050s will bring. TURN TO PAGE 161.

Ugh. Awful hyper-capitalist Martians conquer the planet.

Turns out they've been spending all of the past two decades building advanced war drones and bioweapons. Typical.

They claim all of the earth's best resources as their own and start building tiny mini-utopias at the beautiful, newly restored natural forests and beaches.

At least you've got that shiny badge, though.

- Well, you survived the decade. TURN TO PAGE 194.

We decide that the best way to grow the economic pie for everyone is to reward the people at the top.

There are few economic stimulus packages, but lower tax rates for businesses.

Some company stock prices boom. Unemployment, though, soars. Instead of using the tax cuts to spur job growth, the rich get even richer, while millions line up for food aid all around the world.

In 2028, scientists produce a report saying that despite all the international climate change agreements, greenhouse gas emissions are still rising. Activists propose putting a higher tax on carbon emissions, which companies can offset by doing things like planting trees.

What's your angle on this?

- Are you nuts? Right now we need to *grow* the economy, not raise prices.
  TURN TO PAGE 95.
- That seems like a sensible, free-market approach to solving climate change!
  TURN TO PAGE 258.

The News

All the news, all the time.

3 January 2060

UPDATE YOUR TRACKERS (PAGE 291)

ECONOMY **SWEET** TEMPERATURE **2.3** CONFLICT **SCARY**

TOP ARTICLE

Regional geoengineering by Saudi trillionaires leads to drought and crop failures in India

Farmer suicides at all-time high

OTHER STORIES TODAY

The wealthiest entertainers create utopia bubbles; you're probably not invited

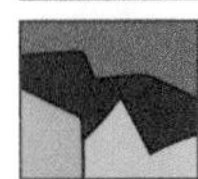

Iyce festival in Greenland fails spectacularly, partygoers stranded as luxury tents sink into melting permafrost

A new company wants to let you grow your own pork... on your own arm?

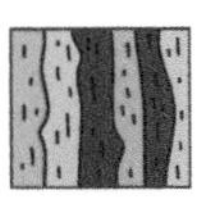

Major landslides destroy Malibu after XÆ A-Xii Musk's cloud-seeding experiments cause unexpected torrential rainfall

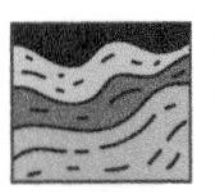

Bangladeshi lowland-elevation project falls through – quite literally

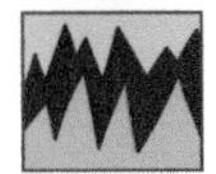

Chinese economy rebounds as geoengineering silver bullet slows rapid warming

Cool, let's see what the 2060s will bring... **TURN TO PAGE 169.**

China buys goodwill by distributing vaccines across Africa and Asia.

Pandemic-ridden countries in South America ask for vaccines, and are told to ask their rich neighbours to the north. The US is angry about this; they've got their own problems and China's just trying to make them look bad.

Organized-crime networks start producing and selling fake vaccines, leading to more scepticism about whether vaccines really work. Families stop speaking to each other over whether or not the risks of vaccination are worth it.

Because of all this arguing, it takes nearly two years until enough of the world is vaccinated that things can return to sort-of normal.

By the late 2020s, economies are battered. Countries need bold new economic-recovery plans. Different groups propose different solutions.

→

Which economic proposal will your news site give the most time to?

- Governments should invest in large-scale public infrastructure projects, like trains and housing. TURN TO PAGE 18.
- Cut taxes and regulations on businesses so they've got more money to grow. TURN TO PAGE 148.
- Our country first! Let's break with our international trade agreements, put tariffs on imports and stimulate our own economy. TURN TO PAGE 224.

The push for universal basic income becomes overwhelming.

Most countries are under pressure to try UBI projects.

There's just one problem: the world's still pretty broke after the disastrous post-Covid 2020s.

> *i* UBI is a government programme in which every citizen, regardless of means, is given a set amount of money on a regular basis to ensure that their basic needs are covered.

How could we fund UBI projects?

- Privatize most social goods like healthcare and education. TURN TO PAGE 24.
- Nationalize natural resource extraction. TURN TO PAGE 264.
- Tax the rich and reform capitalism. TURN TO PAGE 48.

The News

All the news, all the time.

3 January 2030

UPDATE YOUR TRACKERS (PAGE 291)

ECONOMY **HMMM...** TEMPERATURE **1.6** CONFLICT **TENSE**

TOP ARTICLE

Tensions mount over control of Arctic sea routes

Emergency summit aims to prevent new Cold War between US, Canada, China and Russia as melting ice creates lucrative new shipping routes

OTHER STORIES TODAY

Bolsonaro issues new permits for logging in the Amazon after indigenous communities devastated by Covid-19

Bay of Bengal sets new record for number of super-cyclones, millions of people displaced

Matrimonial advertisement: Indian NRI seeks suitable match

Read more **TURN TO PAGE 43.**

Record damages due to floods in the US Midwest are only matched by damage caused by mega-fires on the west coast of North America

What will this new decade bring? **TURN TO PAGE 113.**

There are more desperate discussions about geoengineering.

You've heard rumours around your newsroom that a rich country just wants to try doing something unilaterally, but it would have to be something pretty stealthy so as not to inflame global tensions.

What's your editorial angle on this?

- It's too late to try to get consensus. One country should just take initiative and fix the climate. TURN TO PAGE 80.
- No, we can only do this with the buy-in of everyone. Let's call another global summit. TURN TO PAGE 56.

Over the next few years, the world elects a number of Margaret Thatcher/Ronald Reagan-type pro-capitalist national leaders.

They argue that we've ruined the economy by trying to control it, and we need to let the free market do its job.

They loosen regulations, including environmental laws, which leads to a massive increase in emissions again.

Oh, and all of the new public housing, public transport, schools and hospitals we built in the 2020s? We sell them off to private businesses.

- Well, you made it through the decade! TURN TO PAGE 166.

The News

All the news, all the time.

9 August 2090

UPDATE YOUR TRACKERS (PAGE 291)

ECONOMY **ROCKING!** TEMPERATURE **3** CONFLICT **PEACEFUL**

TOP ARTICLE

New-New Singapore named Best Dome of 2092

"You can't even see the dome," say happy residents

OTHER STORIES TODAY

De-extinction update: Tigers, rhinos latest creatures to be resurrected through gene editing, available to adopt as miniature pets

"There were no domes when my father was president," says William Trump

Seats for simulated Holy Himalayan tourism get booked five years in advance

Looking for an unusual holiday? Try an underwater tour of Miami!

Ten more wonderful things the Council did this year (we love the Council!)

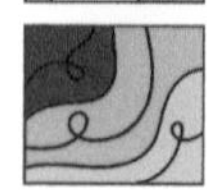

Scientists discuss: "How quickly can we cool the planet safely?"

You made it to the end of the century! **TURN TO PAGE 168.**

Populist leaders emerge calling for us to restrict access to contraception and abortion.

They gain some momentum by appealing to regressive groups who feel like the world has changed too much, too quickly, and win some political power.

But after decades of greater empowerment, women around the world stand together and refuse to accept this.

The movement fizzles out.

- You made it to 2080! TURN TO PAGE 88.

You did not survive the century.

You put too much faith in a technological fix alone solving all of our problems, instead of also doing the urgent, ongoing work needed to change society and stabilize the climate.

To take action TURN TO PAGE 268.

Universal basic income allows many more women to enter politics.

Over the next few years, these progressive leaders pass a number of democratic reforms to make sure that wealth is more fairly distributed.

Overall, people are much richer and better educated.

Climate change is just a technological problem. We'll deal with it later.

What else do people do with their newfound spare time? Well...

There's a new sexual revolution. Birth rates continue to fall, marriage rates plummet, and 60 per cent of women and 45 per cent of men now identify as queer.

What do you think, bigshot editor? Are you okay with this?

- Yes, this is great! TURN TO PAGE 256.
- This is unnatural! Save our families! TURN TO PAGE 83.

The News

All the news, all the time.

9 August 2090

UPDATE YOUR TRACKERS (PAGE 291)

ECONOMY **ROCKING!** TEMPERATURE **2.3** CONFLICT **DANGER!**

TOP ARTICLE

Are you Team Cocoa or Team Lydia?

MicroCorp No-More-Hunger Game down to final two contestants

OTHER STORIES TODAY

"It's worse than I expected, but better than the streets": Part one of our three-part investigative reporting on North America's largest COHO operations (Spamazon)

Trouble in Paradys: Virtual worlds, real targets Read more TURN TO PAGE 15.

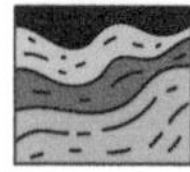

Womb Riot rebels release statement Read more TURN TO PAGE 201.

Thousands apply for one surrogate position: Compensation for surrogacy at all-time low as supply of hopefuls outpaces demand

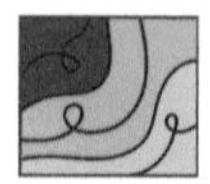

Staying cool in the yard: How cities have adapted to increasing heatwaves with outdoor air conditioning

You made it to the end of the century! TURN TO PAGE 109.

The effects of climate change are getting more and more severe.

Some experts argue that it's time to deploy some form of ambitious solar geoengineering project to buy ourselves time to get the climate back under control. But solar geoengineering technology is still largely untested, and there's a lot of uncertainty about whether it will work, or how it would be managed. Something could go wrong.

We could try solar geoengineering now, or spend a few more years talking about it first, trying to reach international consensus about how we should go about it.

i

Solar geoengineering is a proposed group of technologies that aim to reflect more sunlight away from the earth to cool the climate. These include ideas like injecting sulphur aerosols into the stratosphere, brightening the clouds above the ocean, or even deploying space mirrors. None of these technologies are ready for deployment (in 2021). Research suggests that solar geoengineering could reduce temperatures, but it could also introduce some big new risks, like disrupting rainfall in major food-producing regions. Solar geoengineering has been called a stopgap measure because it does not reduce greenhouse gas emissions, the underlying driver of climate change. It treats the symptoms rather than curing the underlying condition. Solar geoengineering is highly controversial, with some people arguing more research should be done on these

→

technologies, and others saying that tampering even more with our climate system is too risky.

Should we try a last-ditch geoengineering solution?

- Let's keep talking about it for a few more years. TURN TO PAGE 104.
- Let's give it a go. TURN TO PAGE 41.

So, where do we end up by the end of the century?

So close, and yet so far!

We will remember this as the century when we stopped prioritizing growth above all else, and started valuing the truly valuable things: human welfare and the environment. Nature is thriving and species we thought were extinct recover, while new species have evolved for the new wild places of the earth.

But then we lost it all to rich Martians.

How did you get here?

You focused on reducing emissions early in the century, took bold steps to build a more inclusive economy, and prioritized respect for the natural world. Unfortunately, you had no ability to protect the earth you'd built from people who wanted to take it for themselves.

To take action **TURN TO PAGE 268.**

Play with the data behind your decisions at tinyurl.com/mr2kbahw

Phew, okay, we did decide to invest in a planetary clean-up project back in the 2050s.

Luckily that means that there are fewer greenhouse gases in the atmosphere. The planet will warm a bit, but the magnificent Council has enough time to fix the space mirrors before anything too bad happens. Thank you, Council!

One of the potential risks of blocking some of the sun's rays to offset global warming is that if we stopped, the temperature could rise very quickly. This would happen if geoengineering was deployed without society also reducing greenhouse gas emissions at the same time. Suddenly stopping geoengineering would cause temperatures to rebound fast. Scientists call this risk "termination shock" and it would be catastrophic for ecosystems globally.

- Congratulations, you made it to 2080! TURN TO PAGE 103.

Every billionaire has their own pet climate-hacking project.

It's an unregulated mess.

An activist group suggests that the projects should be nationalized and run democratically.

Should we have citizen groups take them over, or keep things as they are?

- Democracy, please! TURN TO PAGE 14.
- Billionaires know best. TURN TO PAGE 112.

The News

All the news, all the time.

3 January 2040

UPDATE YOUR TRACKERS (PAGE 291)

ECONOMY **SWEET** TEMPERATURE **1.7** CONFLICT **SNIPPY**

TOP ARTICLE

Heatwaves drive record-breaking crime rate around the world

Unprecedented rise in domestic violence, assaults linked to impact of rising temperatures on mental health

OTHER STORIES TODAY

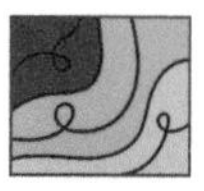

Turmoil as Canada, Russia, and the US threaten to walk out of emergency climate summit in the Maldives

Climate activists stage sit-in at World Economic Forum in Davos, are told to "go home and whine to your dolls" by president of the European Union

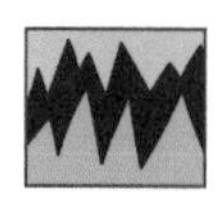

Russian scientists confirm spike in emissions from thawing permafrost, warn warming could cause vicious cycle

Record-low rainfall is a warning sign of Amazon rainforest ecosystem collapse, say alarmed experts

The African Union's Great Green Wall slows Sahara's southward march

On to the 2040s! **TURN TO PAGE 144.**

Uh-oh.

You get a frantic phone call in the middle of the night: a rogue group of democracy activists blew up the space mirror control centre, bringing it offline.

Your lead climate reporter tells you, in a voice choked with fear, that a sudden increase in the temperature could be catastrophic. Our only hope is that back in the 2050s when we launched the space mirrors, we continued to clean up all the greenhouse gases in the atmosphere too.

Oh heck, did we do that?

- If you have the PLAN B badge, TURN TO PAGE 164.
- If you don't, TURN TO PAGE 66.

So, where do we end up by the end of the century?

Humanity is rich and flourishing, but has fewer individual freedoms.

It wasn't an easy ride to get here, though. Many millions of people died or were displaced because of climate change. Over a million animal and plant species have gone extinct. We're now living on a high-tech, fully engineered planet that would be unrecognizable to people of today. We live in the constant fear that if one piece in this complex system fails, that might be the end of humanity.

How did you get here?

You focused on adapting to climate change rather than trying to stop it. You got lucky, and your technological fixes mostly worked (for humans, anyway).

To take action **TURN TO PAGE 268.**

Play with the data behind your decisions at tinyurl.com/2xtj3hy7

We start experiencing even more climate shocks: hurricanes, fires, droughts, flooding, and animal and insect death.

They're coming fast now.

There's a growing movement that calls themselves "Planet B". They argue that geoengineering the earth is a waste of time. We should rather put all of our resources into colonizing Mars on a larger scale.

Should we forget about Earth and put all our hopes in Mars?

- No! Surely it makes more sense to fix the planet we're already on? TURN TO PAGE 47.
- Screw earth, let's all go to Mars. TURN TO PAGE 28.

UPDATE YOUR TRACKERS (PAGE 291)

ECONOMY **YIKES** TEMPERATURE **3** CONFLICT **ALL-OUT BLOODSHED**

TOP ARTICLE

Millions gather in the streets to mark ten-year anniversary since first declaration of war

"We started out calling this the Sky War, but history will know it as the Starvation War"

OTHER STORIES TODAY

Fighting continues as war leaves no clear winner, but plenty of losers

EXCLUSIVE: Leaked transcript of video call between US Leader-for-Life and African President of the Free States

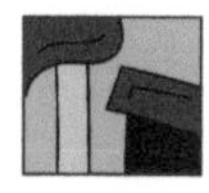

Sea wars and closure of the Suez Canal by Egypt prevent European food aid from reaching drought-stricken India

New victory gardens: Stretch your rations further by growing food at home

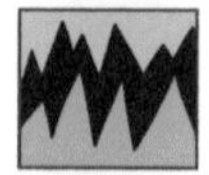

Genetically modified Tibetan rice in high demand as Indian rice yields plummet

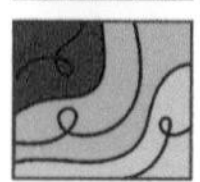

Widespread wildfires destroy fir tree plantations in Greenland, two decades' worth of sequestered carbon go up in smoke

Onwards into the 2070s... **TURN TO PAGE 29.**

Millions try to move to the "green zones", parts of the world that are more liveable.

This leads to a decade of thousands of small-scale regional wars throughout the world.

Oh, hey; quick question: Does anyone have nuclear weapons?

- Oh yeah. They're everywhere. TURN TO PAGE 128.
- No, luckily we disbanded all the nukes earlier in the century – forgot to mention it! TURN TO PAGE 206.

So, where do we end up by the end of the century?

Humanity huddles in the places that are left, poorer, sicker, hungrier, withstanding endless barrages of extreme weather events and plagues, and slowly learning to adapt to them.

Many people join new religions that preach "anti-natalism", the idea that having kids is immoral and it's time for the human race to end.

How did you get here?
You allowed entrenched interests to prevent bold action, and procrastinated until it was too late to change anything.

To take action **TURN TO PAGE 268.**

Play with the data behind your decisions at tinyurl.com/yckzhayk

Mount Agung, a volcano in Bali, erupts, fortuitously cooling the world by 0.2 degrees for the next two years.

Wow, that was lucky.

- You made it to 2070! TURN TO PAGE 247.

One year later, the inventors walk out excited to show off their greatest achievement:

A bubble-wrap mannequin with a wig and lab coat that they've named Henry.

A year is a long time to be stuck in a lab, okay.

No technology to fix climate change, though, sorry.

- Want to rethink that editorial? TURN TO PAGE 20.

The News

All the news, all the time.

3 January 2050

UPDATE YOUR TRACKERS (PAGE 291)

ECONOMY **SO-SO** TEMPERATURE  CONFLICT **FRIENDLY**

TOP ARTICLE

Last harvest of wheat crop in Argentina signals the end of monoculture

Sustainable small-scale organic agroforestry now produces 90 per cent of all food

OTHER STORIES TODAY

"We've turned the tide, but we should expect more disasters in the coming decade," say climate experts, pointing to stabilizing temperatures

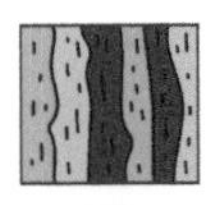

Laugh of the day: Review of a "Green Mall" by a grumpy Gen X shopaholic

Read more **TURN TO PAGE 21.**

Thanks to home insect farms, protein diets have improved significantly

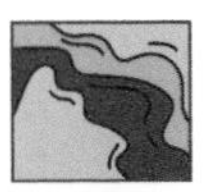

Surface ocean acidification stabilized within safe limits for shellfish

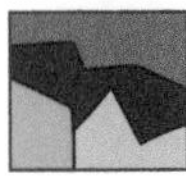

Beavers return to Manhattan as New York City fragments into sustainable communities

Let's see what the 2050s will bring. **TURN TO PAGE 50.**

The world agrees to the fastest-possible transition to a zero-emissions world, even though this will require some changes to people's lifestyles in rich countries.

A lot of people grumble about the lack of alternatives: a lot of things people in rich countries love have been taken away, with nothing offered in their place. Cars, meat and fast fashion soon become luxury items.

We're going to need even more investment into public transport (even in smaller cities), high-quality, affordable protein alternatives, and new ways for people to heat or cool their buildings. Millions of homes will need to be renovated to be more energy efficient, and cities will need to build more renewable energy systems to power everything.

All of this is going to cost a LOT of money.

How should we pay for it?

- Reform capitalism. The super rich should pay for everything, and the new infrastructure should belong to everyone. TURN TO PAGE 48.
- Nationalize the biggest polluters. TURN TO PAGE 136.
- Offer tax breaks and grants to tech companies who are working on solutions. TURN TO PAGE 52.

The Net Zero Emissions Accountability Act is adopted globally by 2025, putting strict limits in place on how much greenhouse gases can be released into the air every year.

The nations of the UN agree that the International Criminal Court should have power to fine and enforce sanctions on anyone who violates these laws.

Bold activists start putting together court cases arguing that historical emitters should pay reparations for loss and damage from climate change. The world declares that knowingly violating pollution laws is a crime against humanity, now called *ecocide*.

By 2029, the biggest emitters have more than halved how much they're polluting the atmosphere, and some CEOs of large fossil-fuel companies have mysteriously gone into hiding. The air is much nicer to breathe all of a sudden, and three million children who would have died from air pollution every year get to grow up.

→

i

The word *ecocide* is an umbrella term for all forms of environmental destruction from deforestation to greenhouse gas emissions. For over 50 years, environmental advocates have championed the idea of creating an international ecocide law that would penalize individuals responsible for destroying the environment. Now in 2021, human-rights lawyers are drafting a definition of the law in the hopes of getting it adopted by the International Criminal Court.

- Congratulations, you made it to 2030! TURN TO PAGE 85.

The News

All the news, all the time.

3 January 2060

UPDATE YOUR TRACKERS (PAGE 291)

ECONOMY **ROCKING!** TEMPERATURE **2** CONFLICT **FRIENDLY**

TOP ARTICLE

People disappointed that they can't choose what weather they get every day

"That's really not how the space fleet works," say bewildered scientists

OTHER STORIES TODAY

Greta Thunberg, age 57, declines leadership role on Global Climate Council: "The young must lead us"

Time to take that Thailand trip at last! Joy as tropics become habitable again

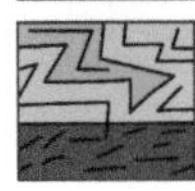

Weaker monsoons in South Asia drive regional water shortages

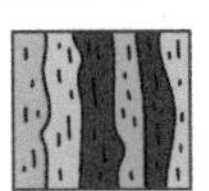

Synthetic fabrics and subdermal implants dazzle at Paris Fashion Week

Read more **TURN TO PAGE 188.**

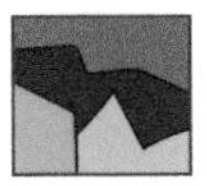

Last hydropower plant closed in Indian Himalayas after successive years of low flows and low electricity generation due to melting of glaciers

Let's see what the 2060s will bring. **TURN TO PAGE 246.**

Economies all around the world have been battered by Covid.

Countries need bold new economic recovery plans. Different groups propose different solutions.

Which economic proposal will your news site give the most time to?

- We need governments to create jobs building new, green public infrastructure like trains and wind farms. TURN TO PAGE 18.
- We need stimulus packages to bail out existing businesses. TURN TO PAGE 200.
- Let's lower taxes on businesses so they invest back into the economy. TURN TO PAGE 148.
- We need a cash injection, fast. Let's sell off new mining and drilling rights. TURN TO PAGE 53.

UPDATE YOUR TRACKERS (PAGE 291)
ECONOMY **NO CHANGE** TEMPERATURE **NO CHANGE** CONFLICT **NO CHANGE**

PRESS RELEASE

Mass abductions hobble India's guest talent programme

BY RAJAT CHAUDHURI

DELHI: Close to a hundred job-seekers from newly flooded areas of neighbouring Bangladesh heading for Delhi were led away at gunpoint on Saturday night by armed intruders who boarded a train in the strife-torn borderland of Uttar Pradesh and Bihar states of northern India. According to Heartland Radio, a local station of the leftist Subah (morning) militia which controls areas south of the Ganges, this is the third mass kidnapping in as many months.

Eyewitnesses told the radio station that the Kalashnikov-wielding kidnappers, who were speaking in a local dialect, boarded the Amity Express around midnight from Dildarnagar and soon began asking

→

passengers by name for their Bangladeshi passports before herding them away towards the pantry car. The intruders got off the train by hacking into the onboard train control system in the abandoned drought-stricken area of Sakaldiha about 50 kilometres west of Varanasi. According to a railway employee, the kidnapped Bangladeshis were packed off in a tourist bus which was heading north towards Ghazipur. Two Indian passengers who provided resistance were shot and are being treated in a Subah militia-controlled hospital.

As a gesture of long-standing ties with climate-ravaged Bangladesh, the Indian government had launched its ambitious Highly Talented Displaced Peoples Programme, HitDip, which offers employment and resettlement opportunities to thousands of professionals and their families in the newly developing smart cities of west and southern India. However with famines and food scarcities plaguing this Asian giant, coupled with unprecedented internal migrations, the programme has come under severe attacks from separatist and other forces who are engaged in pitched battles with a much-weakened government.

A spokesperson for the Subah militia, which acts as an informal law-keeping arm of the provincial government in these far-flung areas, told *UnFake Times*, "Last night's kidnapping seems to be the handiwork of the Ghazipur Five," a dreaded criminal gang. According to Ratan Awasthi, director general of Uttar Pradesh police, "The notorious Ghazipur Five gang is involved in gunrunning and narcotics trade and

→

has been recently involved in kidnapping migrants. They are pushed into slave labour in the opium factories north of the Ganges."

Opium, which was strictly controlled till the early decades of the century, has become a drug of choice in south Asia because of its easy availability. The opium stupor is considered a great escape from all-round despair. Elon Yadav, state government minister of law and justice, said about the kidnap, "These criminals are aided by separatist elements from the north." Vowing to break the nexus between secessionist forces and the drug trade, he requested more central government assistance for purchasing surveillance robots and autonomous weapon systems. "We need this tech to root out anti-national elements and bring the northern areas under the democratic process."

A central team has been rushed to the area.

This article wasn't written or edited by human editors.

Let's see what the 2050s will bring. **TURN TO PAGE 161.**

Vaccine rumours are everywhere.

They're filled with oestrogen and it's a plot to feminize men! They contain microchips so that Bill Gates can track our every movement! They cause cancer so that you're addicted to buying Big Pharma medicines for the rest of your life! Anti-vaxxers hijack vaccine-distribution trucks and destroy millions of doses.

Instead of taking months to roll out the vaccine, it takes four years.

By the late 2020s, economies are battered. Countries need bold new economic-recovery plans. Different groups propose different solutions.

Which economic proposal will your news site give the most time to?

- Governments should invest in large-scale public infrastructure projects, like trains and housing. TURN TO PAGE 18.
- Cut taxes and regulations on businesses so they've got more money to grow. TURN TO PAGE 148.
- Our country first! Let's break with our international trade agreements, put tariffs on imports and stimulate our own economy. TURN TO PAGE 224.

So, what does the world look like at the end of the century?

Looking back, we'd describe the 21st century as one of profound change, mostly for the better.

We will remember this as the century when we stopped prioritizing growth above everything else, and started valuing the truly valuable things: human welfare and the environment.

There's more respect for the idea of environmental stewardship, and we have rewilded a lot of the earth. Most people live in dense, green cities in harmony with nature. There are fewer people in the world and there's much less consumption than in the rich lifestyles of today, but the world is more equal and happier overall.

Nature is thriving and species we thought were extinct recover, while new species have evolved for the new wild places of the earth. We manage to avoid the worst effects of climate change.

How did you get here?
You focused on reducing emissions early in the century, took bold steps to build a →

more inclusive economy, and prioritized respect for the natural world. Shifting to renewable energy and sustainable agriculture in the 2020s was key, but it mattered that you balanced those changes with investing in jobs that support human wellbeing, like those in education and healthcare.

To take action **TURN TO PAGE 268.**

Play with the data behind your decisions at tinyurl.com/mr2kbahw

The Council manages to release pollinating drones and new heat-resistant crops that don't need to be pollinated, so most people are saved.

But there's still a year of food rationing, which makes people pretty miserable.

The Council manages to restabilize the climate, though, once again proving that it is the greatest and humanity can withstand anything.

The weather in most parts of the world is still unreliable. The Council says that this proves that nature can't be trusted, and our best hope is to separate ourselves from it as much as possible.

Our Council invests more money in creating safe climate-controlled cities that are resilient to extreme weather. Some communities start to experiment with building cities that are entirely underground, or are covered with huge domes to protect them from the outside world.

- We made it to 2080! TURN TO PAGE 103.

UPDATE YOUR TRACKERS (PAGE 291)

ECONOMY **NO CHANGE** TEMPERATURE **NO CHANGE** CONFLICT **NO CHANGE**

PRESS RELEASE

Paris Fashion Week 2060 special report

BY SOPHIA AL-MARIA

It's been a devastating few decades for the fashion industry as it transitions away from the traditional "seasons" of S/S and F/W to favour a timeless Resort format more befitting the sunny new outlook of the 2060s. With the newly launched space mirrors climate-controlling our lives to the minutest degree, we are seeing more and more reflective inspirations from bias-cut rescue-blanket gowns and party attire to disco ball inspired legwarmers and

→

McQueen-esque armadillo heels that pay homage to the mirrors dreamed up and launched from Shanghai's labs last year.

In the collections? A kaleidoscopic medley, iridescent PU fabrics and an attention to the unseen spectrums of ultraviolet and infrared. It's an odd mash-up in terms of garments, with light basic T-shirts and trousers trimmed with synthetic furs (microscopic metal fibres that read to the eye as a grey fur) – furkinis, for example. In other words, it's glamour, a return to the skimp and flair of some 70 years ago. Iridium-infused polymers create trippy subtleties, and leave the texture of clothing distinctly metallic.

We never thought we'd see the sun set on overengineered doomer fashions of the thirties, the apocalypse we anticipated was actually a psychic fatigue of it. To some, though, the glamour we have now isn't one of choice, so much as luxury's assault – unceasing and perfectly acclimatized to our consumption before we ever know it. The more avant-garde designers are trying to break out quite literally from the civil enclaves of perpetual resort and pamper. Silhouettes mock the glamorous revivalism found today, with floor-length dresses designed to trail and soil at the hem, exposing the steel-wool-ish makeup that most textiles have nowadays. It's rebellious, a little Romantic, and a bit dirty. People have called it juvenile and fantastic in equal measure. Other designers have taken their fabric shears to entire cultures – a designer's final look proposes a hijab-cum-qipao halterback, of course debuting to mixed applause.

→

China's hold on luxury is still unmatched. Some insiders noted the lineage from the highly engineered styles that came maybe only a decade or so before; fabrics like synthetic calluses that came to define the era. Tongue-in-cheek screenprinted silicon shirts, suits and little black dresses leaned into the trompe l'oeil and contemporary fascination of daily life's make-up not being what it seems; the 16k screen prints sat atop opaque and warm-to-the-touch synthetic vegan leathers that are still ever so popular.

Even further, outerwear (that is, to the extent to which a garment can be said to be definitively "outer" anymore) tends to be imbued with any properties a person might need, as points of difference – that of probiotic goodness, energy, analytics and drug administration. However, woven fabrics are not exactly extinct; cut any of these membranous materials open (if you can) and you will be met with an explosive tuft of fibres. They're the foundational base patterns by which all clothes are still made, giving all necessary ranges of movement and near-impenetrable durability. We have the aeronautic and defence sectors to thank for this one, with our current landscape of fibre-reinforced PU latexes, silicons and leathers being the new petro-infused norm.

Even with all of this defiant chemical romance to the garments, there's an odd but not at all unexpected return to nature, a rewilding of our minds and art – one of the few places this exists. For example, skirt silhouettes are cloven, with the midi and mini variations giving way to the "chidi", or orchid-like imbricate forms of petals: hemlines often

→

appear both high and low simultaneously, depending on who you ask – think McQueen bumsters or Deroyli hot sqorts.

With that said, some of the most expensive garments out there are single-use, degrading over the course of a day (32 hours is the current record) – we have Korean designers and K-Beauty brands operating from Rwanda to thank for this. It's all in stark opposition to fast fashion that is made more-or-less on demand and from scratch, though takes an epoch to decompose naturally. Turnover is high, but this, today, is offset by apps that make the redistribution of these clothes as seamless and as easy as anything. Clothing is both re-gifted and repurposed into heat blankets, shelters and rafts. These humanitarian-aid and crisis start-ups have kickstarted what's known as the donation economy, giving the fashion industry and its consumers a smug sense of purpose. In many ways it's because high fashion exists between the permanent and temporary, amid a ballooning middle class. Everything is dropped in fortnightly slots of edits, rather than "collections". New studies into the importance of news cycles' effects on human psychology reveal that news is a kind of food for our anthropocentric social disposition. Withdrawal from people and news is understood to be fatal in many cases – all press really is good press when it comes to staying alive.

There are, still, overarching trends that stick within these brief windows. Body modifications have made faces and other aspects of human presentation identikit among the mainstream – the generalized feline features once attributed to camera iPhone apps like the "aesthetic →

analytics" giant Facetune have actually been grafted onto people's faces – one needn't honour their genetics seeing as it's so cheap and accessible. There are trends tending towards subdermal implants (both tech-based or aesthetic) and scarification of flowers, animals and corporate brand logos taking the place that tattooing once had. More rebellious youths are opting for eye tattoos, fangs and nail-bed implants which have become perhaps the breakthrough beauty innovation of our time; perma-nails have become the new luxury accessory and jewellery item to have. Made in any material from ceramic to titanium, with cheaper variants made with highly durable polys. The biggest brands market them like fine jewellery for women and top-end watches for men, but have found more practical everyday use in the place that wallets and purses once held.

It must be said, it feels like we're at the technological bleeding edge, with much excitement to be expected with what we wear in the coming year – it's all about leaning into the reflection of ourselves, not away from it.

Let's see what the 2070s will bring. **TURN TO PAGE 246.**

Uh, really? Okay...

You publish an editorial saying that we should bet on a technological breakthrough to magically fix our problems. It's a very convincing piece!

World leaders lock ten of the smartest inventors on Earth in a room for a year and ask them to invent a technology to magically fix climate change.

They warn you there's a one-in-six chance that they'll find something.

Roll a die and see if we get lucky.

- If you roll 5 or less, TURN TO PAGE 174.
- Wow, you rolled a 6? Okay! TURN TO PAGE 100.

The News

All the news, all the time.

9 August 2090

UPDATE YOUR TRACKERS (PAGE 291)

ECONOMY **ROCKING** TEMPERATURE **1.7** CONFLICT **RUN!**

TOP ARTICLE

Are you Team Cocoa or Team Lydia?

Final two contestants compete in talent show hoping to win spot in Martian-controlled oasis

OTHER STORIES TODAY

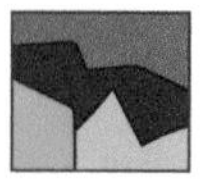

Solar-powered Cape-to-Cairo magnetic levitation train "derails" mid-journey in the Congo basin, causing severe impacts to the home of mountain gorillas and creating mass outcry across the world

Thousands apply for one surrogate position: Compensation for surrogacy at all-time low as supply of hopefuls outpaces demand

Outcry as logging begins in Amazon rainforest for the first time since 2030s

You made it to the end of the century! **TURN TO PAGE 163.**

The world agrees!

And decides that the global climate fund can also be used to support any projects that increase GHW (global human welfare) anywhere in the world.

Global inequality plunges to the lowest level it's ever been. People are generally well fed, well educated, happy and healthy. There's not a lot of luxury consumption, but people are able to spend time building strong, healthy relationships with each other.

We're really starting to see the benefits of having restored wild space. Animal populations are recovering, and the world temperature has started to drop slightly. Scientists say we've turned the corner.

It's not over yet, though. Coastal communities are still being hit by extreme weather events and flooding.

- Woohoo! You made it to 2060. TURN TO PAGE 265.

UPDATE YOUR TRACKERS (PAGE 291)
ECONOMY **NO CHANGE** TEMPERATURE **NO CHANGE** CONFLICT **NO CHANGE**

PRESS RELEASE

California passes ban on selling and trading animal products: Boom in illicit underground meat restaurants

BY MARIA TURTSCHANINOFF

The state-wide Californian raid of illegal meat restaurants, commonly known as “meat-ups”, as well as importers of animal flesh for human consumption, has raised criticism in the US as well as abroad.

“We uphold that a total ban of meat is not the way to go,” said Autumn Peltier, prime minister of Canada. “We absolutely need to radically cut back on the consumption of meat, but my programme has been one of incentives, taxes and leading by example. A ban always leads to the trade and consumption going underground as well as profiteering.”

A total of 59 restaurants were raided on Friday night, and over 150 people were arrested, proprietors and staff alike. This publication has

→

the names of actors, athletes and politicians who were frequenting these establishments Friday last. We choose not to publish them, as we feel that might encourage some of the loudest protestors of the ban. Police had been given instructions to let the customers go, and to only arrest the staff, as it is not illegal to eat meat, only to produce, smuggle or sell it.

During the last era of prohibition, in the 1930s, the illegal speakeasies are said to have been elegant places where the demi-monde gathered to play games, dine and dance. But this reporter accompanied a raid to a small house in West Covina where the air was heavy with the greasy smell of fried steaks and burgers. On the tables were paper plates, plastic cutlery and bottles of beer and wine. A man in his mid-sixties, presumably the proprietor, tried to protest that this was a gathering of family members, not a restaurant, and that eating meat was not a crime. This was an obvious falsehood, as the people gathered, approximately 24 people of all ages and ethnicities but predominantly male, could not, when pressed, give satisfactory explanations of how they were in fact related. "We're cousins" did not impress the detective in charge.

"But we are not so much after the restaurants, or 'meat-easies,'" said Governor Christina Schwarzenegger. "The bigger issue is the illegal import of beef and pork from other states."

What will this new decade bring? **TURN TO PAGE 249.**

The terrorist groups are rounded up and sent to re-education camps.

Your newspaper submits a request to send a journalist in to report on what's happening in there. Nobody gets back to you.

- Well, you survived the decade. TURN TO PAGE 156.

The News

All the news, all the time.

3 January 2040

UPDATE YOUR TRACKERS (PAGE 291)

ECONOMY **HMMM...** TEMPERATURE **1.8** CONFLICT **SCARY**

TOP ARTICLE

World Trade Organization breaks up

China leads belt-and-road countries to form new trading bloc

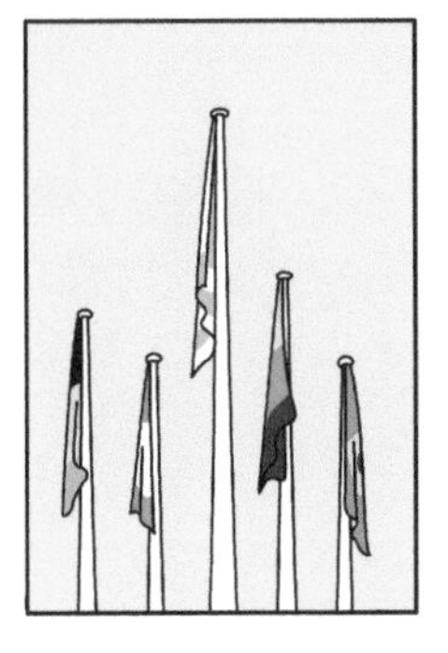

OTHER STORIES TODAY

Japan and Australia deny asylum to boats of migrants seeking new homes after fisheries collapse from marine heatwaves in the Indo-Pacific

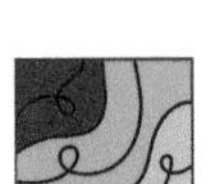

African elephants listed as critically endangered as industrialization of agriculture fuels conflict with farmers

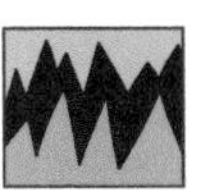

Reinsurance companies file for bankruptcy as Arabian Sea cyclones batter Indian and East African coastlines

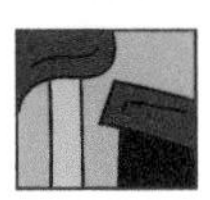

Forced relocation from Sundarbans sparks widespread violence in eastern India

On to the 2040s! **TURN TO PAGE 254.**

We choose to focus on saving existing industries (including dirty ones).

A lot of stimulus money ends up with banks, automakers, fossil fuel companies and so on, without any incentives for them to change how they do business.

This does prop up the economy... at least, it looks like it does on charts of the stock market.

A lot of people are still out of work, though, and struggling.

- You made it to 2030. TURN TO PAGE 95.

UPDATE YOUR TRACKERS (PAGE 291)
ECONOMY **NO CHANGE** TEMPERATURE **NO CHANGE** CONFLICT **NO CHANGE**

PRESS RELEASE

Statement from Womb Riot demanding radical eco-political reforms now

Thanks to our reporter Lauren Beukes for bringing this to our attention.

BY LAUREN BEUKES

We are Womb Riot, a group of 18 women and non-binary people with wombs, who have been working as human incubators for the wealthy of Lovelace Two, the fifth-largest colony state on Mars, and the most exclusive.

Our unborn charges include the heirs to some of the greatest tech dynasties, including the granddaughter of Dylan Okereke-Clifford, and the triplets of X Æ and his husband, Li Ma. As of this morning at 05h47, we have barricaded ourselves in the aerial suite of the Ambani Stadium floating above Ascraeus Mons.

The Deimos Dervishes will not be playing their scheduled game against the New York Yankees today. Nor will the rich parents who have plundered our planet have their developing foetuses safely returned to

→

them. Not until they are willing to take action, to instigate meaningful, grand-scale change. Now.

Like our colleagues who serve as organ hosts (COHOs), we have lived our lives as pampered human nests under terrible strictures that infringe on our human rights, our freedom of movement, and our freedom of choice. To those who say we chose to put ourselves in this position to be exploited, we say an unjust society offers no choice. We live in a world where billions barely survive in abject poverty amid desertification and rising seas and temperatures, pummelled by unending natural disasters.

The rich have taken our future from us. We are only returning the favour.

Our demands are simple, as they are old as time. We demand equality for all, the right to clean water, food, safety, security, bodily autonomy, and the chance to pursue happiness for ourselves and our children, without having to surrender ourselves to the whims of the wealthy and privileged few.

What does this mean in practice?

We demand that the tech dynasties and conglomerates acknowledge the anger of the people and immediately cease the violent suppression of the X-tinction Riots playing out across Earth.

We demand the immediate release of X-tinction leaders including but not limited to Charne Allais, Kim Ahn and D'Angelo Greenwood, who are

→

being held in inhumane conditions in Jakarta, Medellín, Detroit and beyond.

We demand that the tech dynasties surrender their assets equal to 80 per cent of their accumulated wealth and that this money be immediately redistributed for ecological reparations under the guidance of X-tinction and their various NGO partners as per the Accra Papers of 2088.

We demand that Teslamerica shuts down the A-93 SRM mirror that has been climate-privileging North America.

If these demands are not met, we will start to terminate these pregnancies, one at a time.

We know they will be coming for us; special forces and private security armed with non-lethal weapons they hope will subdue us, and lethal ones for when that proves futile.

This is why we have chosen to place ourselves in the world's eye, in the glass VIP box of the Ambani Stadium, accessible only by air. We are not afraid. We want you to see what they are capable of, how far they will go. We want you to see that we will not stand down. For all our futures.

We are Womb Riot. We demand change.

You made it to the end of the century! **TURN TO PAGE 109.**

The world is now capturing more greenhouse gases than it produces, but we're still recovering from damage that was done to the environment earlier in the century.

In 2062, we get unlucky. Hot oceans lead to a series of terrible hurricanes across the southern US and Caribbean, destroying thousands of homes and ripping up forests.

We're already capturing a lot of greenhouse gases through natural methods like planting trees and preserving wild areas, but some groups say we need to speed things up with high-tech carbon removal technology.

i

If we were to reduce all greenhouse gas emissions to zero in 2021, temperatures would probably stabilize within the next 10 to 30 years. But some parts of the climate system respond much more slowly than temperatures. Past greenhouse gas emissions have "locked-in" an amount of glacier and ice sheet melting. That doesn't mean our actions don't matter: every 0.1 of a degree of extra warming translates into more lives lost and more species extinct.

→

What do you think we should do?

- Let's stick to doing things the natural way, thanks. TURN TO PAGE 110.
- Let's invest in carbon removal technology. TURN TO PAGE 12.

Phew.

There are a lot of small-scale conflicts and civil wars over resources. Luckily, they never escalate into global or nuclear war.

- Okay, at least most of us are alive. TURN TO PAGE 140.

The News

All the news, all the time.

9 August 2090

UPDATE YOUR TRACKERS (PAGE 291)

ECONOMY **DISASTER** TEMPERATURE **3.7** CONFLICT 

TOP ARTICLE

Nuclear winter causes first snowfall in decades in the Swiss Alps

Water reservoirs finally replenished as people rush to the ponds despite high radiation warnings

OTHER STORIES TODAY

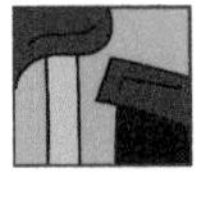

Two-headed wolves slaughtered by doomsday cult, the Volcanists, who believe a major volcanic eruption will cool the planet

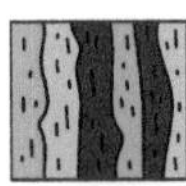

Art review: The Tate Twilight

Read more **TURN TO PAGE 242.**

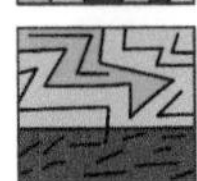

Efforts to curb superflu transmission in New Delhi fail; the New Race cult detonates a nuclear device in the epicentre of the pandemic

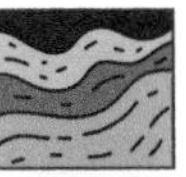

Restaurant review: The Blue Queen of DC

Read more **TURN TO PAGE 250.**

Submerged cities of the world form union, declare independence

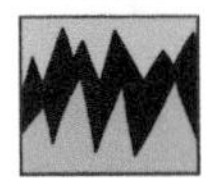

Newborn deformity and cancer incidence rates steadily increase after a decade of nuclear detonations

You made it to the end of the century! **TURN TO PAGE 27.**

Your readers are convinced by this argument.

It's important to move slowly when tackling something as important as climate change!

By 2028, the world's economy has rebounded to pre-Covid levels. People feel flush. There's a boom in luxury consumption, construction, and international travel. Global greenhouse gas emissions haven't really reduced much, despite the investment in new green infrastructure.

Award yourself the NICE SWEATPANTS badge (page 289).

- You made it to 2030. TURN TO PAGE 95.

For most of the century, news organizations like yours have been the voice of moderate centrism, cautioning people against bold action.

Your readers have started to feel like that attitude is part of the reason the world is so difficult now. Readers have even had the audacity to write in and say that you are part of the problem!

Weirdly, you are becoming much less influential than you used to be.

- Rebel groups ignore you and hack the climate anyway. TURN TO PAGE 139.

The world has been far too busy worrying about war to do much about climate change. It's getting hotter, which means more droughts, more hurricanes, and more wildfires. The worst impacts are happening in poor countries.

The leaders of 50 countries in the Global South (low-income countries) band together and say *enough*. Stop bickering! We have to stop all emissions within the next 15 years, and urgently try to reverse the damage that's already been done. Rich countries, who've caused most of the damage, need to pay to help low-income countries adapt to the changed climate.

Most of your readers live in rich countries. Will you encourage them to take responsibility and get serious about reversing the damage we've done to the planet?

- The rich world should absolutely stop polluting, but every country needs to find its own solutions. TURN TO PAGE 64.
- Ha! No way. That sounds like a plot to weaken our country! TURN TO PAGE 40.
- What if the Global South made them agree? TURN TO PAGE 253.

UPDATE YOUR TRACKERS (PAGE 291)
ECONOMY **NO CHANGE** TEMPERATURE **NO CHANGE** CONFLICT **NO CHANGE**

PRESS RELEASE

Trust is stronger than blood: An interview with Babylondon mayor Lena Lenin

Lena Lenin talks Marie Antoinette, the myth of the nuclear family and the power of exotic stimulants.

BY SOPHIA AL-MARIA

When Lena Lenin was in Year 5 at primary school in what was then Elephant and Castle, south London, a New Eden officer gave a talk to their class. She told the students that they had the potential to do great things in spite of the planetary hardship their generation would endure. They only had to work hard and dream big. Thirty years later and Lenin has done both of those things. The New Eden project that originated with her team when

→

she was a young councillor in the borough that would become "the bubble" we see thriving today as it welcomes migrants and others displaced by climate disasters. And while the dome is being built to rival all previous civil-engineering projects in the world, Lena Lenin is still dreaming of what might be possible for the people of this old town and also what could be possible for those who decide to ignore government guidance and stay outside the biospheres being built around the globe today.

Born in a council estate, Lena Lenin was raised by a queer co-op that shared a floor inside the Skyton Terrace Estate, a sprawling warren of public housing completed in the 1970s at the height of what would later be known as the 20th century's greatest folly: relying on middle-aged white men to predict and build the future. These "chambers of despair", as Skyton was once described by a former MP from the area, are what prodded Lenin to seek change, heading up solar panelling for the council at an early age and pushing through legislation to allow her friends and neighbours (all of whom she counts as family) more agency and autonomy from the city government and the metro police, both of whom seemed both historically and in the early half of the century to actively want the inhabitants not only disenfranchised but in many cases dead.

But how things have changed. Now in the year 2080, Lenin is the mayor of a world-building project the likes of which few civilizations have ever known. And hers is a vision of universal basic income, of radically non-judgemental policies regarding everything from gender →

and sexuality and sexual health to drug use as well as spirituality, and all of this is funded by the new release of petro-dollars from the UK's allies in Nigeria and Colombia and the Gulf Oil states.

We sat to have a chat with Lenin over her favourite new designer drug, a Guatemalan bird-picked snuff Membre power, and strawberry macaroons from one of the few antique heritage suppliers that will be continuing to work within the new Babylondon economic bubble zone (BEBZ): La Duree.

You've been keen to include certain "heritage" companies in the transition period. These macaroons are delicious, but why French pastry when it has been such a symbol for your detractors who say you have a latter-day Marie Antoinette attitude in spite of your humble beginnings?

"Let they/them eat cake" was an early jab my enemies liked to throw at me. And so I decided I wanted to just own it. The ideals of the French Revolution haven't died in the 21st century. We believe in many of the Jacobin ideals for "the people"; we could just do without the terror. The brutality and ugliness and puritanism of its methods have left their warning. I want there to be joy and beauty in the future. I believe there will be both. And sometimes it's the little things. Like a small sweet pastry that makes a big difference in morale. So yes, I thought it would be good to have a bit of old-world refinement alongside the new-world tech and sweeping legal and social changes we are implementing. No

→

police and everyone gets a couture stipend and as much free time as they like to pursue their interests and joys as long as it doesn't impinge or infringe on anyone else's. With that in mind, we chose a few special products and companies to represent the old world. Think of BEBZ on closure in 30 years' time as having been a sort of curated corporate ark if you will. We're keeping Burberry but getting rid of Bentley, for example. No need for us to be driving long distance inside the BEBZ and let's use those petrochemicals for other purposes like keeping our servers chilled. Amiright?

You grew up in a co-op. Many people mourn the loss of nuclear families. Do you think your background means you have less empathy for those who miss the old blood way of kin-making?

For me trust is stronger than blood. Any day. If I don't trust you, I won't turn to you in a crisis. If we share the same bloodfather, there is no reason I should turn to you for help. The kin I choose and that any one of us chooses on this planet are our true love and saviours. And that rarely comes in a heteronormative package. This is why my campaigning and life's work has been to normalize what used to be considered deviant. That includes everything from identifying as asexual to having access to assisted suicide. I run on a radically non-judgemental platform. You can do what you want as long as you don't infringe on

→

anyone else's ability to do what they want. It's simple. It's natural. It's really really real.

Hardline anti-BEBZ campaigners have latched onto your open and public drug use and pro-synth policies in the face of GMO scandals as a sign you are not in a fit state of mind to lead. What do you have to say in response to that?

Do you recall when we used to call them Chads? There will always be a threat that these nutjobs will resurface, but for the most part they exist in the outer limits of chat rooms and live outside the bubbles. We have anti-xenophobic failsafes in place for future dreamers who want to join us, however there will always be a bad apple in a basket of home-grown fruit and that's why we retain one ugly element of the old world: the death penalty.

I'm sorry but some might say you're dodging the question about drug use, Mayor Lenin.

On the contrary. If in the old days, they were comfortable being led by the sober and senile rather than the compos mentis and high, then I ask you to consider simply where that led us. We've got to try something new. And if de-moralizing the social order and taking the stigma out of the

→

things traditionally and wrongly believed to be "bad" doesn't lead to an immediate and radical change for the "good" then I don't know what will.

That is all.

Just one more decade to go. **TURN TO PAGE 39.**

An unlikely alliance of Saudi billionaires who miss being able to go outside without dying, disgruntled former SpaceX engineers, and K-Pop stars looking to leave a legacy surprise the world by launching a fleet of space mirrors to reflect sunlight away from the planet.

A K-Pop supergroup promotes the project by releasing a song called "Space Space Planet Baby" that tops the charts in every country.

The project is successful beyond anyone's expectations. And because the K-Pop stars promote the project, no one can be too mad about it.

- So K-Pop saves the world? TURN TO PAGE 248.

Rich countries refuse!

The Global South enters a strike. Those countries stay united and cut off natural resources. They refuse to export food, minerals or metals until the rich world agrees to their demands.

This hits the rich world where it hurts. Their manufacturing industries can't produce much, so the new iPhone launch is delayed. Worse, no one can get their hands on a decent cup of coffee.

- The rich world caves after six months. TURN TO PAGE 195.

The News

All the news, all the time.

3 January 2040

UPDATE YOUR TRACKERS (PAGE 291)

ECONOMY **SO-SO** TEMPERATURE **1.6** CONFLICT **SNIPPY**

TOP ARTICLE

Lagos's living buildings win Eco-Innovation of the Year prize

Inspired by indigenous building methods, bacteria-filled walls trap carbon and heal themselves when damaged

OTHER STORIES TODAY

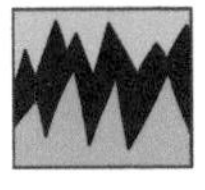

Third consecutive year of net-zero greenhouse gas emissions mean we're on track to reach peak temperature within next decade, say very relieved climate scientists

Southern Florida rewilding project launched as extreme events make the region uninhabitable (for all but Florida men)

Wangari Maathai Institute launches virtual university, offers training in eco-feminist natural resource management

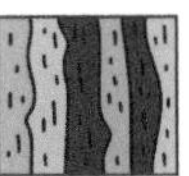

Indian Ocean and Pacific islanders resettled in new host countries as sea ice melts and oceans rise

Saudi Arabian women overthrow the monarchy as eco-utopian movement sweeps across Middle East

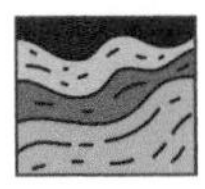

Nigella Lawson, age 79, issues cookbook with grow-your-own meat recipes

Everybody's feeling optimistic about the 2040s! **TURN TO PAGE 142.**

A new technological breakthrough means we've become aware of huge new oil and gas reserves in the Arctic.

Thanks to global warming, they're going to be easier to exploit than ever before!

Greenland is a poor island, so they have no choice but to offer licences to the throngs of companies that flock there looking to make quick money. This entirely reverses any progress that's been made.

Somehow, despite everything we've tried, we're polluting even more than ever before.

- You made it to 2050. TURN TO PAGE 82.

Climate change disrupts wildlife migration, leading to the emergence of a terrible new virus, popularly known as Wisconsin flu.

Within six months, it spreads all around the world, killing two per cent of the world's population and decimating the global economy.

Talking about geoengineering is nobody's priority right now.

- Well, you made it through the 2060s. TURN TO PAGE 226.

The News
All the news, all the time.

3 January 2070

UPDATE YOUR TRACKERS (PAGE 291)

ECONOMY **NO CHANGE** TEMPERATURE **NO CHANGE** CONFLICT **NO CHANGE**

PRESS RELEASE

Obituaries
Capitalism: 1760–2068

BY MARIA TURTSCHANINOFF

Unnoticed by most, Capitalism has passed away at the ripe old age of 308 years. Its death throes during the last decades were violent and harvested many casualties, but in the end Capitalism went with a whimper, not a bang.

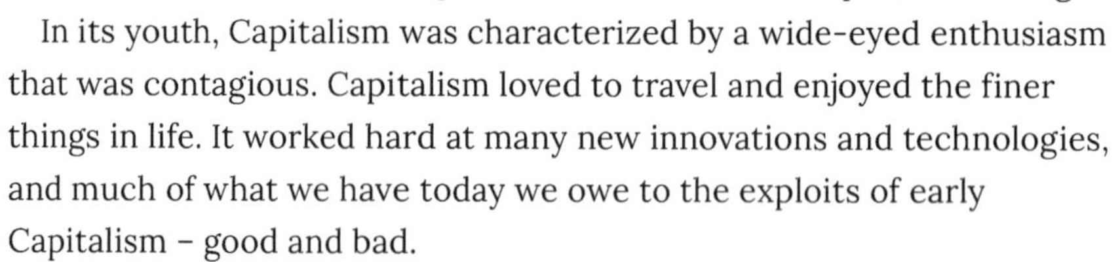

In its youth, Capitalism was characterized by a wide-eyed enthusiasm that was contagious. Capitalism loved to travel and enjoyed the finer things in life. It worked hard at many new innovations and technologies, and much of what we have today we owe to the exploits of early Capitalism – good and bad.

Capitalism had a certain charm that was hard to resist. Even its enemies admit as much. There were many people who suffered the evils of it, yet fought for it, tooth and nail.

"It took me a long time to see Capitalism was abusive," says a former lover who wishes to remain anonymous. "When times were good, they

→

were really good. You know? There were many small warning signs, but I chose to ignore them. You tell yourself that it's not that bad. My kids tried to warn me, several times, that Capitalism wasn't good for me. But I was completely caught up in the race to reach for a little more, always more, not caring who I stepped on in the process. I didn't see that the one I was hurting the most was myself. It wasn't until I finally dumped Capitalism that I started seeing all the ways in which I had suffered."

The former lover is now in a healthy relationship with Ecosocialism. "It's not perfect, because nothing is. But we work at it together, every day. I am so much happier now that I know I am not hurting those around me – human and non-human societies alike."

Capitalism did not take gracefully to being rejected. In the last decades, many of its cronies tried to fight innovations like employee representation on company boards in the 30s and the replacement of GDP with GHW (global human welfare) in the 40s. Capitalism tried to keep its hold on its staunchest allies by filling them with fear of the new world that was emerging, which led to a lot of unrest and revolts in many places globally. Many lost their lives as a result. Including nature in human measures of success was one of the hardest pills an ageing Capitalism had to swallow.

As happens with many who live long lives, Capitalism had, in the end, outlived most of its friends. The last of them left for Mars in 2062.

Onwards into the 2070s! **TURN TO PAGE 114.**

Countries start to pass protectionist laws and break trade agreements.

The US–China trade war escalates, leading to a number of shortages.

Warmer weather is causing ice to melt in the Arctic Ocean. Canada, the US, China and Russia now have a new northern border to protect.

Countries quietly start to divert post-Covid economic stimulus money into beefing up their militaries. You know, just in case.

> *i* As sea ice melts, shipping lanes are opening up in the Arctic, making it an increasingly profitable region. But this is causing geopolitical tensions, as countries scramble to secure their northern borders. Already, governments have sent their militaries to the region in order to ensure that their countries' stakes in the Arctic are protected.

- You made it to 2030. TURN TO PAGE 153.

Migrants are welcomed as "human capital" and resettled in new, high-tech cities all around the world.

We're spending more money on education and healthcare than ever before. A host of new technologies are developed that help us to make the new cities more resilient against climate shocks.

- You made it to 2050. TURN TO PAGE 59.

The News

All the news, all the time.

3 January 2070

UPDATE YOUR TRACKERS (PAGE 291)

ECONOMY **NOT GREAT** TEMPERATURE **2.75** CONFLICT **SNIPPY**

TOP ARTICLE

President of European Commission: "Stop blaming America for the Wisconsin flu"

"Our changing climate is why that raccoon met that bat in the first place, which means we're all at fault"

OTHER STORIES TODAY

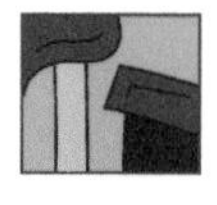

Fears of conflict, counter-geoengineering deter China from deploying solar geoengineering; "another decade lost" say dejected activists

Global day of mourning declared for the millions dead from Wisconsin flu

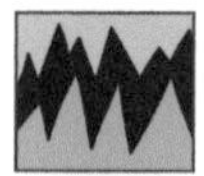

Booming megacities prompt African continental free-trade zone to negotiate trade deal with Brazil for food

Amazon basin formally loses "rainforest" categorization, reclassified as "savanna"

Onwards into the 2070s! **TURN TO PAGE 261.**

Phew, is it just me, or is it getting pretty hot in here?

The world is really starting to feel the effects of climate change. 2041 is a particularly brutal year: so many disasters happen all at once that many of them don't even make it onto the front page of your news site.

The disasters displace millions of people. In the US, droughts and hurricanes force people north. Many of them lose their homes and livelihoods. Bangladesh has become largely uninhabitable due to flooding. Failing fish stocks in the tropics lead to near-starvation among communities that depend on fishing to survive, and force people into cities. Plans have been made for large parts of Hong Kong, Miami, Lagos and Manhattan to be abandoned to the rising sea water.

R.E.M.'s classic song "It's the End of the World as We Know It (and I Feel Fine)" is the most listened-to song of the year.

Millions want to resettle in more protected places. Most resettle in their own country, but many also cross borders in an attempt to give their families a better life.

i

As climate change leads to rising sea levels and extreme events like repeated heatwaves, droughts and floods, many parts of the world are becoming harder to live in, forcing people to leave their homes. Over a billion people live in areas exposed to land degradation or climate-related sea-level rise. Most of this migration will happen within countries, not between countries.

→

How should society respond?

- There are just too many people to assimilate all of them. Let's build refugee camps in unpopulated areas. TURN TO PAGE 105.
- Let's make it easier for businesses to sponsor working visas, so people can only move to a country if they have a job there. TURN TO PAGE 234.
- Refugees are valuable, especially in rich countries with ageing populations. Welcome them and build new housing for them in cities. TURN TO PAGE 101.

Large swathes of the world are declared "orange zones".

These are places where food no longer grows, temperatures are uncomfortable for most people, and there are frequent disasters.

What remains of the governments of these countries are flat broke.

What should we do with the orange zones?

- Let multinationals buy them for cheap. TURN TO PAGE 73.
- Strong countries should claim them and build strategic military bases on them. TURN TO PAGE 102.
- Agree to leave them be. TURN TO PAGE 51.

We keep trying to reduce pollution through small, incremental changes.

We slightly increase taxes on greenhouse gas emissions, but there are a lot of loopholes in these laws, and businesses generally find ways to get around changing.

We are polluting less than we were in the 2020s, but we're far off the goals scientists told us we needed to reach to prevent catastrophic climate change.

Hopefully that won't be a problem or anything.

- Well, you made it through the decade! TURN TO PAGE 166.

The News

All the news, all the time.

3 January 2070

UPDATE YOUR TRACKERS (PAGE 291)

ECONOMY **THIS IS NICE** TEMPERATURE **2.3** CONFLICT **SCARY**

TOP ARTICLE

Blue Ivy Carter's Project Sparklecloud creates cool Connecticut oasis

Housing prices skyrocket in cooler states as tech and entertainment businesses flee drought-ridden California

OTHER STORIES TODAY

"They're using people like petri dishes": Anonymous employee from largest vaccine manufacturer publishes shocking open letter

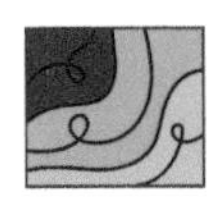

So you want to go to Mars: Getting there when you don't have the cash

Read more **TURN TO PAGE 31.**

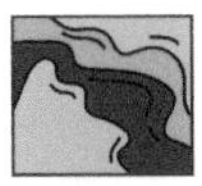

The way things were: Norwegian art installation features holograms of whales returning to the ocean

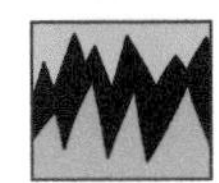

Researchers point to increasing ocean acidification as cause of mass ocean die-offs: "All these projects to reflect the sun haven't done anything to address the chemicals we've poured into the atmosphere"

Onwards into the 2070s! **TURN TO PAGE 165.**

Climate rebels around the world start implementing small-scale projects to hack the climate.

These projects somewhat work, but have unintended consequences. Communities have to adapt to a lot of sudden weather changes: areas that had a lot of rainfall suddenly have to cope with droughts; cities that had become more resilient against droughts suddenly have to deal with flooding.

The world erupts in a series of violent local conflicts over suddenly scarce resources. You publish a report saying that three-quarters of the people in the world now live in countries experiencing civil wars.

The effects are terrible. Millions starve. Billions are displaced by war. This decade is called "the terrible 70s".

Award yourself the I SURVIVED THE 70s badge (page 289).

- When will this decade end? TURN TO PAGE 111.

The EU and US claim that the protests are being stirred up and funded by the Chinese government in an attempt to undermine their democracies.

They claim that it's all being triggered by a targeted disinformation campaign on the hottest new social network: Spldge.

Funny... China, which is seeing protests of its own, claims the same thing.

And no one has a good explanation for why the protests are happening simultaneously in most countries in the world.

Everyone is angry. Nothing is done about climate change.

- Well, you made it through the decade! TURN TO PAGE 199.

A lot of the migrants are offered "corporate citizenship".

Business visas mean getting a job with one of the big multinationals is the only way migrants are allowed to live in one of the parts of the world that are still easy to live in. Losing their job means exile.

> *i*
>
> As climate impacts worsen, many people will be displaced from their homes. Adapting to climate change in a just and equitable way will require us to rethink how we accommodate displaced people. Climate justice is migrant justice.

- And exile is a problem, because large parts of the world are becoming unliveable. TURN TO PAGE 229.

The News

All the news, all the time.

5 January 2080

UPDATE YOUR TRACKERS (PAGE 291)

ECONOMY **SWEET** TEMPERATURE **2.2** CONFLICT **SCARY**

TOP ARTICLE

The new "dick-tators"

Millions in the streets as Coalition for Democracy protestors rally against TeslaTown

OTHER STORIES TODAY

Banks warn that protestors caught by facial-recognition AI risk damaging their social-credit scores, will be denied home loans and healthcare

The rise of the super wives: How to find the perfect woman who'll give you that edge in the boardroom

Still having it all? Ratio of women in the workplace continues to dwindle as more families need to supplement diet with homegrown veggies

Last acre of Amazon rainforest privatized to provide carbon offsets

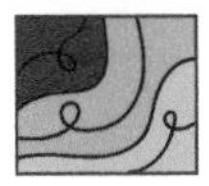

Miss blue skies? New augmented reality eye implants let you choose whatever sky colour you like!

Just a couple more decades to go. **TURN TO PAGE 259.**

You publish an editorial arguing that we need to try a geo-engineering solution. But what will that solution look like?

Roll a die!

- If you roll a 1, TURN TO PAGE 80.
- If you roll a 2 or higher, TURN TO PAGE 248.

We respond by investing even more in healthcare and social support systems.

This generation will be able to keep working (and live better lives) for longer.

- You made it to 2080! TURN TO PAGE 88.

Your news site writes off the movement as being just a bunch of kids getting themselves tied up in knots over nothing, encouraged by vandals and hooligans who just want an excuse to loot stores.

After months of regular protests, when there is still no response from leaders, the activists eventually give up and the protests die down.

- Well, you made it through the decade! TURN TO PAGE 166.

The News

All the news, all the time.

5 January 2080

UPDATE YOUR TRACKERS (PAGE 291)

ECONOMY **HMMM...** TEMPERATURE **3.1** CONFLICT **TENSE**

TOP ARTICLE

Rebel climate hacking causes droughts, crop failures across northern hemisphere

Southern hemisphere agriculture revives for the first time in decades

OTHER STORIES TODAY

International rebel alliance led by Nigerian guerillas dredge up decades-old strategy from Twitter archives, launch distributed solar geoengineering in southern hemisphere to green the Sahel

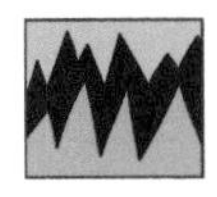

Switzerland bans avocado exports to US, demands lower prices on Alaskan rice

Legal battles continue over Moscow's attempt to expel climate refugees from Siberian bread basket

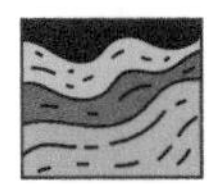

Remaining American airlines file for bankruptcy following heatwave-induced flight cancellations

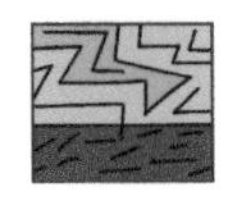

The "year without monsoons" leads to record-breaking meningitis epidemic in Mali

Just a couple more decades to go. **TURN TO PAGE 81.**

We will never surrender! The war goes on.

Six months later, retaliating to US threats to "pull another Hiroshima", Japan releases an engineered blight that wipes out the world's maize crops.

Farmers scramble to produce alternative crops, but the whole world's food-supply chains can't adapt fast enough. Most countries introduce rations to try to share food fairly, but in some places, there just is not enough, and people starve.

That year, millions of people die of starvation, mostly across the Americas.

The war continues.

- Well, you made it to 2070. TURN TO PAGE 170.

The world launches a global climate fund.

A taxation pool that taxes those who produce pollution or consume luxury goods, and uses this money to build renewable energy systems and fund indigenous communities and co-operatives around the world who act as stewards of the environment, which act as carbon sinks.

Global South countries band together and argue that they shouldn't have to contribute to this pool, and also that old IMF loans should be forgiven, because of the historical debt of colonialism.

Should rich countries agree to this?

- That seems fair. TURN TO PAGE 195.
- Absolutely not! TURN TO PAGE 218.

UPDATE YOUR TRACKERS (PAGE 291)
ECONOMY **NO CHANGE** TEMPERATURE **NO CHANGE** CONFLICT **NO CHANGE**

PRESS RELEASE

The Tate Twilight

BY RAJAT CHAUDHURI

Open daily: 7 p.m. to 6 a.m.

The Geiger-counter-wielding guards scan you through into Delhi's abandoned central metro station, now the venue of Tate's newly opened Twilight exhibition space. The main concourse has VR interactive walkthroughs of doomsday scenarios, visualizations of a shutdown of the thermohaline circulation, or you can pose questions to the Buddha delivering his first sermon. It's also possible to chat with a Mayan soothsayer, marvel at Pieter Bruegel the Elder's "Triumph of Death" or behold a lifelike Kali, the dark goddess of time with her garland of human heads. More familiar terrain includes spying upon terrorists nuking the Great Gustaff sulphur dioxide tower, rolling on the deck of an Arabian Sea oil rig as a

→

supercyclone hits, or chatting with a mind-uploaded simulation of corporate guru Jack Windows or Greta Thunberg.

Most of the tunnel-halls are dedicated to the 21st century and you begin with a virtual swim along London's subterranean Fleet river. Encounter floating artefacts and objects, like the severed head of an assassinated Jewish revolutionary who requests a drink while narrating how they are fighting against gene-tweaked Russian oligarchs with heat-tolerant skin, who now lord over a partially submerged London that seceded from the UK.

From there, venture into the Chinese-controlled uranium mines of an overheated Niger. Listen to a clandestinely videoed 2080 account of a white Frenchwoman engaged as slave labour who is suffering from radiation-induced lung cancer. The holotext details how these mines lit up the homes of France till a three-way tussle between the French, Chinese and militants broke Niger apart.

Segue to a largely abandoned Paris where soaring temperatures have sent the Ministry of Environment into the catacombs. Among the objects on display in this hall are a laser guillotine and a compassion suit: voluntary suicide equipment with a supply of helium. A holovision of the perpetual secretary of the Académie Française recites an experimental prose poem extolling the virtues of a helium-laden departure.

Head east. Swim along a waterworld colony of Hong Kong's Kowloon with its expensive undersea bubble homes and floating shanties packed →

with Pacific islanders. Stroll along extensive sea walls built by American POWs and Uighur detainees till you bump into a real cha chaan teng serving noodles and steaming-hot milk tea. Free, if you listen to a lecture about the importance of freedom.

The Tate Twilight offers an honest and gut-wrenching, intense experience, peppered with the dark hilarity of our times. Try on Japanese air-conditioned hotpants with self-cleaning nanotech, a mid-century craze fuelled by a super-scorcher summer, or watch the results of a singular attempt at de-extinction in the form of a climate-friendly Bengal tiger which refuses meat and is no bigger than a cat.

Closer to the exit gates are softly-lit caverns. Here a solarpunk community activist weaves gentle notes on her dotara while others offer free permaculture classes and video game renditions of Callenbach's *Ecotopia*. Within the digital walkthrough of a multispecies Himalayan eco-town stand men in grey personal protective garments wielding megacorp vaccine shots... You pays your money and you takes your choice.

One minor quibble: It seemed the core VR platform running the walkthroughs of the Tate Twilight had been hacked because all through the tour we kept encountering phantasmic characters, one with a yellow mop and another with cold beady eyes, whispering in our ears, urging us to buy smart weapons and seabed methane futures.

Tip: Carry a torch, there will be power cuts.

You made it to the end of the century! **TURN TO PAGE 27.**

Confronting the fact that national leaders aren't going to do enough, María Carlini, the newly elected 27-year-old mayor of Mexico City, calls a World Cities Summit, inviting the mayors of the largest cities from all around the world.

Over 150 city leaders assemble, together representing a billion people, and one-third of the world's economy. They realize that they don't need to wait for their national leaders to act.

The cities commit to the 2035-Zero goal: for the combined emissions of all of these citizens to be zero by 2035. This goal is measured collectively – cities that had a head start in transitioning to renewable energy invest in technology to capture greenhouse gases, to offset the emissions by cities that start later. The cities work together, pooling their resources and sharing information about how to change.

These changes immediately make life better for the inhabitants of these cities. Air pollution clears up, there are waves of new jobs in green tech, and there are more green spaces. Megacities become even more desirable places to live, and more people move to them.

There's still a lot of inequality, though. What should we do about that?

- Fix capitalism! TURN TO PAGE 60.
- Inequality's fine as long as it spurs innovation. TURN TO PAGE 52.

The space fleet becomes the target of attack by rogue states.

They try to hack it to benefit their own microclimates.

How should we deal with these rogue actors?

- Give the global climate council an army. TURN TO PAGE 121.
- Tell them they are being very naughty and politely ask them to stop messing with the space mirrors. TURN TO PAGE 266.

The News

All the news, all the time.

3 January 2070

UPDATE YOUR TRACKERS (PAGE 291)

ECONOMY **THIS IS NICE** TEMPERATURE **1.5** CONFLICT **PEACEFUL**

TOP ARTICLE

Five-day festival declared for restoration of Lake Chad

After shrinking to one-tenth of its full size in the 2020s, Lake Chad has returned to the size it was in 1960

OTHER STORIES TODAY

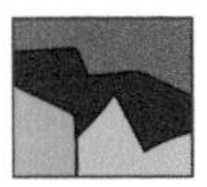

Berlin conference decolonized: Communities given back access to land despite national borders

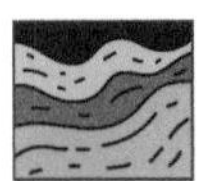

Obituaries: Capitalism (1760–2068)

Read more **TURN TO PAGE 222.**

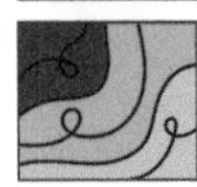

World's last billionaire leaves for Mars: "It might not have an atmosphere, but at least you won't try to tax my hard-earned money!" says woman who inherited 90 per cent of her wealth

Global permaculture awards go to Peruvian farmer, Juan Ruiz, who has resuscitated an indigenous potato cultivar not seen for the past century

Farmers in Australia diversify back to sheep and cattle – not just camels

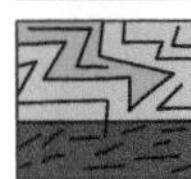

Monsoon season is back: Celebrations across the subcontinent

Onwards into the 2070s! **TURN TO PAGE 114.**

We successfully launch a fleet of space mirrors that reflects some sunlight away from the planet, reducing warming.

In a historic meeting, we establish a new global climate council that oversees the project, and begin to deploy it. All council members must be under the age of 20, because they have the most stake in the future of the planet.

The council is warned that if the space-mirror project ever fails, there might be terrible consequences. The world can also invest in a backup plan, building machines that suck carbon out of the atmosphere, but it will be expensive. The cost will mean a five per cent reduction of everyone's UBI (universal basic income), which will be unpopular!

Should we also invest in the planetary clean-up project?

- Nah, we trust the space mirrors! TURN TO PAGE 179.
- Yes, let's make sure there's a plan B. Award yourself the PLAN B badge (PAGE 289), then: TURN TO PAGE 179.

The northern hemisphere's summer of 2031 is particularly brutal:

Hurricanes and wildfires batter the US while hundreds die in a heatwave across southern Europe and central Asia.

In November 2031, world leaders come together for another climate conference. We've gained some momentum on climate change, but clearly a lot more needs to be done.

Every world leader reads your news site, so tell them. How aggressive should we be about reducing greenhouse gas emissions?

- There's no time to waste. Let's commit to net zero emissions by 2035, regardless of how hard it is. TURN TO PAGE 176.
- Let's prioritize switching over to renewable energy, but not do anything that's going to interfere with people's lifestyles. TURN TO PAGE 245.

PRESS RELEASE

The Blue Queen of DC: A review

BY RAJAT CHAUDHURI

Food reporter

Arriving here is no hassle. An armed Italian usher in a gondola rows up to greet you at the rusting jetty across the submerged Indian Head highway. He offers a round-bottom flask of golden tej. You enjoy sips of the honey wine while the boat heads mid-river to the stilted reincarnation of the old National Harbour, now a fin-de-siècle maelstrom of gene-hackers, doomsday cults and highlife holdouts like the Blue Queen – DC's last Ethiopian eatery.

As you are winched up from the water by a platoon of Pacific islanders, you cannot miss the tanks of the star-shaped hydrogen plant looming like a kraken in the rippling waters of the Potomac. The

→

restaurant is run as a co-operative by the secretive Volcanist cult, and the rustic chic interiors have 19th-century chromolithographs on the walls and long-spouted black clay coffee pots in the alcoves. The warm glow of old-fashioned lamps and traditional basketwork tables for communal repasts impart a homely air; an underwater section offers more familiarity, with mood-matched shape-shifting chairs and holographic projections of the giant obelisks of Aksum.

The dark-eyed Abyssinian owner speaks Amharic, English and Chinese but her American all-male staff, mostly out-of-work farmhands, need language synthesizers. Service is quick; the waiters know their menu and they will diligently explain how they strike a balance between authenticity and "compulsions of the age".

If you have booked early which, considering volatile circumstances, means at least 60 days prior, expect to be fed well. The omnipotent injera bread is soft, spongy and baked with care from first-generation GM teff which comes closest to the lost original. I still harbour a sinful fascination for real beef, and their minced-meat kitfo, which I ordered rare or lebleb, was tender and melt-in-the-mouth delicious, accented by the chilli burst of mitmita seasoning.

Half of my folks are fish-loving Bengalis from the sunken deltaic region of eastern India, which is why I couldn't resist the seductive asa tibs. The fried two-headed perch (guaranteed radiation-free) was boosted with the peppery and aromatic berbere spice blend, while my

→

bodyguard went for juicy lamb tibs in the sinful red awaze sauce with its spiced-up heart. But the meat came from a lab.

Vegetarians can try their luck with a variety of stews (wots) prepared with red lentils, chickpeas and more, but most of this would be printed and flavoured airfood proteins rustled up by microbes from carbon dioxide. There are shady gene-hacked varieties too, most popular being the skin-darkening tikil gomen cabbage stew. Round it all off with the full-bodied Ethiopian coffee.

The Volcanists believe a series of eruptions will cool the planet, buying enough time for the forces of justice and reason to prevail. Our taste buds teased and bellies full, as the human-operated lift dropped us back into the swollen Potomac, we could only hope they know what they're talking about.

You made it to the end of the century! **TURN TO PAGE 27.**

The Global South enters a strike.

Those countries stay united and cut off natural resources. They refuse to export food, minerals or metals until the rich world stops polluting and agrees to contribute to a global environmental repair fund.

This hits the rich world where it hurts. The manufacturing industries can't produce much. The new iPhone launch is delayed. Worse, no one in the rich world can get their hands on a decent cup of coffee.

- The rich world caves after six months. TURN TO PAGE 37.

Phew, is it just me, or is it getting pretty hot in here?

The world is really starting to feel the effects of climate change. 2041 is a particularly brutal year: so many disasters happen all at once that many of them don't even make it onto the front page of your news site.

The disasters displace millions of people. In the US, droughts and hurricanes force people north. Many of them lose their homes and livelihoods. Bangladesh has become largely uninhabitable due to constant flooding. Failing fish stocks in the tropics lead to near-starvation among communities that depend on fishing to survive, and force people into cities. Plans have been made for large parts of Hong Kong, Miami, Lagos and Manhattan to be abandoned to the rising sea water.

R.E.M.'s classic song "It's the End of the World as We Know It (and I Feel Fine)" is the most listened-to song of the year.

Millions want to resettle in more protected places. Most resettle in their own country, but many also cross borders in an attempt to give their families a better life.

i

As climate change leads to rising sea levels and extreme events like repeated heatwaves, droughts and floods, many parts of the world are becoming harder to live in, forcing people to leave their homes. Over a billion people live in areas

→

exposed to land degradation or climate-related sea-level rise. Most of this migration will happen within countries, not between countries.

How should society respond?

- There are just too many people to assimilate all of them. Let's build refugee camps in unpopulated areas. TURN TO PAGE 105.
- Every country should have to accept a number of refugees proportional to their historical carbon emissions. TURN TO PAGE 129.
- Let's create a new visa scheme allowing refugees to resettle if they take a job building green infrastructure. TURN TO PAGE 234.
- Send them back. Migrants should be settled internally, somewhere else in their own country. TURN TO PAGE 87.
- Refugees are valuable, especially in rich countries with ageing populations. Welcome them and build new housing for them in cities. TURN TO PAGE 101.

The world is really starting to feel the effects of climate change.

The popular view is that it's far too late to stop any of this from happening: we have to *adapt* to the new climate.

Some cities manage to do this quite well: we build huge barriers to keep back the ocean, we develop new food sources, we build huge machines to filter pollution from the air so it's breathable. With our new wealth and stable society, it feels like climate change is something we can manage.

But in 2041, we are particularly unlucky. It's a brutal year. So many disasters happen all at once that many of them don't even make it onto the front page of your news site, and they happen too fast for us to address them.

The disasters displace millions of people. In the US, droughts and hurricanes force people north. Many of them lose their homes and livelihoods. Bangladesh has become largely uninhabitable due to flooding. Failing fish stocks in the tropics lead to near-starvation among communities that depend on fishing to survive, and force people into cities. Plans have been made for large parts of Hong Kong, Miami, Lagos and Manhattan to be abandoned to the rising sea water.

R.E.M.'s classic song "It's the End of the World as We Know It (and I Feel Fine)" is the most listened-to song of the year.

Millions want to resettle in more protected places. Most resettle in

→

their own country, but many also cross borders in an attempt to give their families a better life.

> *i*
>
> As climate change leads to rising sea levels and extreme events like repeated heatwaves, droughts and floods, many parts of the world are becoming harder to live in, forcing people to leave their homes. Over a billion people live in areas exposed to land degradation or climate-related sea-level rise. Most of this migration will happen within countries, not between countries.

What's your angle on this?

- Refugees are valuable, especially in rich countries with ageing populations. Let's welcome them into our cities. TURN TO PAGE 225.
- These immigrants are coming for your UBI! TURN TO PAGE 57.

The News

All the news, all the time.

3 January 2030

UPDATE YOUR TRACKERS (PAGE 291)

ECONOMY **THIS IS NICE** TEMPERATURE **1.6** CONFLICT **SNIPPY**

TOP ARTICLE

Brazil builds giant statue of Jeff Bezos hoping to win lucrative mining contract

Amazon looks for new rare minerals supplier for world's largest solar panel project, humanitarians raise concerns

OTHER STORIES TODAY

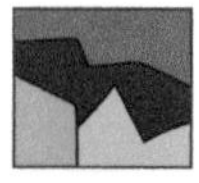

The rising dead: The clean-up crew that's dealing with more than 1,000 thawing bodies on Everest's Rainbow Road

Vaccine Queen opens the first of 100 tech universities across West Africa

Read more TURN TO PAGE 78.

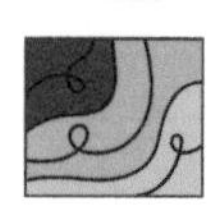

Teenage inventor of cloud-brightening tech becomes world's youngest billionaire

"We are Boston strong": City devastated by Category-5 hurricane rebuilds amid the rubble

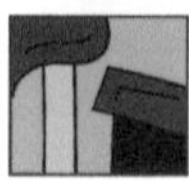

Shell builds thousands of carbon capture machines across Australian outback: "It's cheaper to buy offsets than to reduce emissions," admits CEO

What will this new decade bring? **TURN TO PAGE 11.**

Our attempts to hack the climate start to work, and create gorgeous, liveable micro-paradise zones in parts of the planet.

You have to be very rich to live there, though.

And everyone else, kept out of these paradise zones, forced to struggle against an increasingly hostile environment? They are angry.

Billionaires start a reality show where lucky outsiders compete for limited spots in their new paradises. The huge underclass fights over these reality show spots, or the few decent jobs, like being a pregnancy surrogate (or an organ host – part of a concentrated organ-harvesting operation – COHO) for billionaires.

- Well, you survived the decade. TURN TO PAGE 160.

Sorry, this isn't a democracy anymore. Public opinion doesn't matter.

The terrorist groups are rounded up and sent to re-education camps.

- Well, you survived the decade. TURN TO PAGE 156.

Feeling increasingly hopeless, various groups of climate rebels get together and start discussing cheap ways to hack the climate.

There are risks.

Should we try a last-ditch geoengineering solution?

- Why not? We have to try something. TURN TO PAGE 46.
- It's too dangerous. TURN TO PAGE 209.

The News
All the news, all the time.

3 January 2070

UPDATE YOUR TRACKERS (PAGE 291)

ECONOMY **ROCKING!** TEMPERATURE **1.5** CONFLICT **PEACEFUL**

TOP ARTICLE

The Global Climate Council is great!

They saved our planet, now they can fix all of our other problems

OTHER STORIES TODAY

Malaria reappears in Nigeria, but no problem because the Council deploys a vaccine in record time

Quality of life skyrockets in once-poorer countries; "Sovereign states led to so much injustice," says new minister for global fairness

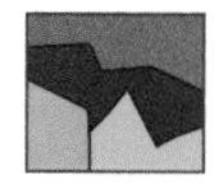

Young people in northern hemisphere experience first cold winter: "Thanks, but we hate it. Can we get global warming back?"

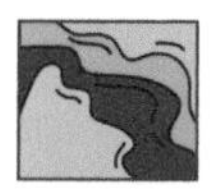

After 20 years, sea ice is spotted again in the Arctic; ice sheets keep collapsing, but Council spokesperson assures us they will become more stable in time

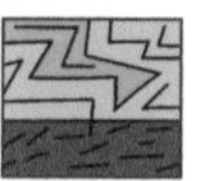

India sees their first season without monsoon due to disruption of the hydrological cycle

The trees are coming! Increased woody vegetation expansion across Africa turns grasslands into savannas and savannas into forests

Onwards into the 2070s! **TURN TO PAGE 167.**

We relax housing regulations and provide generous tax breaks to businesses building new housing complexes in the cities.

A number of successful housing conglomerates pop up. They build a lot of luxury high-tech buildings for the super-rich. Middle-income families are forced to take out exorbitant loans to afford substandard housing, while lower-income families are pushed out into crowded slums at the edges of cities.

- It's still better to live in the cities than outside them, you hear. TURN TO PAGE 229.

We realize that there's one clear way we could fund UBI *projects: fracking.*

We can consider natural gas a transitional energy source that helps to wean us off dirtier technologies.

Your paper, as an influential voice, supports this. Right now, it's more important that we build a more equal economy than worry about pollution.

Award yourself the UNIVERSAL BASIC INCOME badge (page 289)!

- UBI projects are rolled out in most of the world. TURN TO PAGE 96.

UPDATE YOUR TRACKERS (PAGE 291)

ECONOMY **THIS IS NICE** TEMPERATURE **1.6** CONFLICT **FRIENDLY**

TOP ARTICLE

Indigenous peoples and local communities (IPLCs) take their seat at the UN

"About bloody time," says Māori elder

OTHER STORIES TODAY

Bonfire of the passports: Thousands of protestors say national borders should be abolished

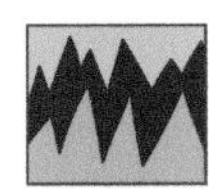

Unhappy capitalist billionaires plan Mars colony, hoping to be free from "the tyranny of taxation"

Abandoned skyscrapers become home to indoor guerilla farmer revolution

Scotland announces plan to shut world's last remaining golf course and rewild the land

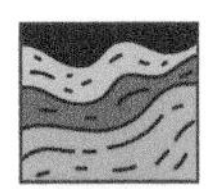

Locust outbreaks not seen since 2020 leave many asking, "Are they delicious?"

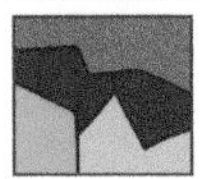

All vertebrate wildlife population movements to be tracked by satellite to measure "pulse of the planet" score

Let's see what the 2060s will bring. **TURN TO PAGE 204.**

Yeah... that doesn't work.

The rogue states successfully hack the space mirrors and bring down the whole system.

It turns out that we'd become more dependent on geoengineering than we realized. As soon as we stop, the temperature shoots up very suddenly.

> *i* One of the potential risks of blocking some of the sun's rays to offset global warming is that if we stopped, the temperature could rise very quickly. This would happen if geoengineering was deployed without society also reducing greenhouse gas emissions at the same time. Suddenly stopping geoengineering would cause temperatures to rebound fast. Scientists call this risk "termination shock" and it would be catastrophic for ecosystems globally.

- This is BAD. TURN TO PAGE 221.

We let multinationals buy up all this land.

Out of sight of regulators, they restart polluting extraction of the natural resources in these areas and build terrible work camps.

- You made it to 2050. TURN TO PAGE 146.

Thanks for playing!

We hope you've enjoyed thinking about what the rest of the century might look like, and that you're inspired to have some influence in the real world.

The climate has already changed, and it will continue to change. But that doesn't mean it's over.

Imagine you're driving on a highway. You start at a world that's 1°C hotter than pre-industrial times. 1°C isn't paradise, but it's a pretty good place to live. But the further you drive from 1°C, the more difficult things get for us. We're already driving on the highway.

The highway goes from here all the way to 4°C by the end of the century. 4°C sucks. We do not want to live there. 4°C doesn't mean that the whole human race ends, but it does mean that we will have to withstand more disasters, more disease, more fights over scarce resources. Millions of people and animals will die, and life will be harder for most people. We want to get off the highway long before we get to 4°C.

Right now, in 2022, we're coming up to the offramp that goes to 1.5°C. It is still plausible that we could take bold action now and get off the highway here. 1.5°C is a worse neighbourhood than 1°C, but it's MUCH better than 4°C. We could adapt to 1.5°C. We could flourish.

And yes, at some point, we will have driven past the 1.5°C exit (we haven't yet, but we're getting close and haven't even pulled into the slow lane yet). But when we drive past the 1.5°C exit, should we just throw our

→

hands up and say "oh well, I guess we're going to 4°C then". No! We should try to get off at 1.6°C. Or 1.7°C. Or 1.8°C. Every 0.1°C matters, and we will have to keep fighting over each fraction of a degree. And at every single one of those exits, there are twisty side roads that lead to vastly different worlds. Worlds that are more or less equal, worlds with more or less biodiversity, worlds with more or less responsive governments and corporations and media machines. As well as fighting to limit degrees of temperature rise, efforts to address climate change are about making the future that we and those who come after us will share.

Climate change is made and experienced by people. How the future will play out is going to be shaped by choices taken by people today, and tomorrow, and the day after.

Our choices matter. It's not over. There are many choices still ahead of us.

Now, make a choice in the real world.

- For some discussion questions, TURN TO PAGE 276.
- Learn more about the climate emergency. TURN TO PAGE 278.
- Find a climate justice group near you by visiting 350.org.
- Read about how we made this book. TURN TO PAGE 279.
- Play again! TURN TO PAGE 10.

Read the stories

→

Playing as a group

Survive the Century has been used in classrooms and conference rooms around the world. There are two main approaches we've seen work well.

THE DISCUSSION VERSION

45-60 minutes

1. Divide your group into teams of 3-4.
2. Direct everyone to survivethecentury.net (or give them a copy of this book) and give them a few minutes to play through the first decade (the 2020s) in their team, stopping when they get to the first newspaper screen. Encourage them to hit the "back" button and try out at least one different choice.
3. Lead a 5-minute discussion as a whole group, based on the discussion questions (page 276).
4. Continue like this through the remaining decades.

If pressed for time, you can allow teams to play the whole game through to the end (15 minutes) before leading a 15 minute group discussion. The game can also be played individually as homework, followed by a 20-30 minute discussion in class.

→

THE DEBATE VERSION

50-70 *minutes*

1. Divide your group into teams of 2-4. In groups of 5 people or fewer, each person works individually.
2. Assign each team or individual a newspaper section from the list below. Explain that you (the facilitator) are the editor of the newspaper, and the teams represent subeditors who must try to convince you to make decisions that will appeal to their readership. For instance, the Business section wants to prioritise the interests of business people. Tell the participants that bribery, threats, lies and other creative forms of rhetoric are 100% encouraged.
3. Project survivethecentury.net up on a screen where it's visible to the whole group. Have the facilitator read the prompt and choices out loud.
4. Allow the groups a couple of minutes to discuss what decision they will support, and then give them a chance to pitch it to you. Allow the teams to have a small debate back and forth. The facilitator should pick the most convincing argument.
5. Continue through the game. You can put every decision up for debate, or only crucial ones, depending how much time you have.
6. Leave 20 minutes of discussion time at the end to talk about what happened (out of character). What were the most compelling

→

arguments made? What did this exercise show you about how climate politics happen in the real world?

Newspaper sections

- Business
- Lifestyle and Culture
- Celebrity Gossip
- Environment
- Technology
- International Politics
- Local News
- Religion
- Youth
- Sports
- Obituaries
- Humour

ASSIGNMENT IDEAS

- **Individual:** Write a short news article from the future reflecting on how the changing climate might have impacted a specific area of human culture. It can be a straightforward news article, or an unusual format like a travelogue, agony aunt column, quiz, restaurant review, or the transcript of a political speech.
- **Group:** Write and perform a short news or variety talk-show programme from the future that features celebrity interviews, live-reporting, or advertisement.
- **Individual (advanced):** As the Survive the Century team, we sense-checked our temperature pathways using a climate modelling tool called En-Roads (available for free at en-roads.climateinteractive.org). Use this tool to create a dashboard that simulates the world in 2100 at 1.5°C, and at 3°C. Write up a timeline of political and social choices that might have led to that outcome.

Discussion questions

- When you read the story, what choices seemed to matter the most?
- Was your first instinct to try to fix things, or unleash your inner supervillain? Why?
- What feelings came up for you? Why?
- In reality, who's making choices about the world's response to the climate emergency? Why do you think they're making the choices they're making?
- How do normal people who care about the world convince powerful people to make different choices?
- What will climate change look like in your hometown? How might it affect your local economy, environment, politics and communities?
- Have you already directly experienced one of the effects of climate change?
- How are you already seeing people in your community adapting to the effects of climate change? What are some adaptations you might expect to see in the future? What are the consequences of those adaptations?
- How old will you be 20 years from now? Go back to the news headlines from 2040 (see page 99, page 199, page 55, page 219 and

→

page 166). How do you expect the world to be different when you're that age?

- What do you think this story gets wrong? (We'd love to hear your thoughts: hello@survivethecentury.net)
- Did the fictional articles help make some potential climate choices feel more plausible?
- What did you learn about how the changing climate might impact society, politics, technology, the economy, and vice versa?
- What are some positive side effects that could come out of the world's response to the climate emergency?
- Let yourself be optimistic about politics for a moment. Imagine that the world makes all the positive changes you hope for and we're living in the best version of 2100 that you can imagine. What would you love about being part of that future?
- Will anyone benefit from global warming?
- What are the reasons to be hopeful about the future?
- What actions can you take today, in the real world, to be part of building the future you want?

Recommended resources

Want to learn more about the climate emergency? Here are some of our favourite resources.

- *How to Save a Planet* (podcast): gimletmedia.com/shows/howtosaveaplanet
- *All We Can Save* (essay anthology), Ayana Elizabeth Johnson and Katharine K Wilkinson
- *This Changes Everything*, Naomi Klein (nonfiction book)
- *Ministry for the Future*, Kim Stanley Robinson (novel)
- *Braiding Sweetgrass*, Robin Wall Kimmerer (nonfiction book)
- *The Great Derangement*, Amitav Ghosh (nonfiction book)
- The Kurzgesagt video series on climate change (search YouTube)
- *Our Planet* starring David Attenborough (documentary series)

About

Survive the Century grew out of a workshop convened in 2019 by **Dr. Christopher Trisos** and associate professor **Dr. Simon Nicholson**, climate researchers at the University of Cape Town and American University, respectively. This workshop, under the umbrella of the US-based National Socio-Environmental Synthesis Centre (SESYNC), brought together people from different backgrounds from around the world to think about the climate emergency. **Sam Beckbessinger** was one of those attendees and the game's co-creator, and is responsible for all of the bad jokes in this story.

This story is a work of fiction, but it is informed by real science. Over 25 scientists, researchers, policymakers and subject-matter experts were consulted in putting this story together. Sci-fi writers from Finland, India, Qatar and South Africa helped craft the storylines and come up with the hilarious headlines. You can find a full list of their names on the credits page (page 284).

WHY WRITE A BRANCHING STORY ABOUT CLIMATE CHANGE?

Climate scientists are often pushed to the front of the climate movement. They were among the first to raise the alarm about what's in store if we don't slow down climate breakdown. Their cutting-edge predictions – ice melt, species extinctions, extreme drought and wildfires – are the best way to communicate what's really at stake, so the theory goes.

But sometimes, the sophisticated computer models climate scientists create are an insufficient guide.

STORIES ARE POWERFUL

The possible climate futures that await us will be a big, complex mess of political and social forces interacting with environmental change. That's hard to model. Of course, we still need excellent scientific models to help us understand what the future of the planet might look like. But we also need other tools.

This is where fiction comes in. In recent years, we've seen the rise of climate fiction, or 'cli-fi' – a sub-genre of speculative fiction that tells the stories of what might be to come as our climate changes beyond recognition.

This book is an experiment in the power of fiction to help people vividly imagine a range of climate futures lying ahead.

Do you call school strikers heroes or hooligans? As the climate crisis displaces millions of people, do you back corporate visas, harsh refugee camps, or resettlement in richer countries? The story confronts the reality that all of us wield a certain kind of power – one that deals in

→

stories, not the scientific method. And it brings to life the fact that the stories we tell to ourselves and others have very real consequences for the future.

BEYOND TEMPERATURE

When the future of the planet is already so extraordinary, why turn to fiction?

First, it's how humans work. We make sense of our past, present and future by telling each other stories: narratives with a beginning and end, and with something important at stake. That fills an emotional gap – engaging heart as well as head – that no amount of research evidence can. Stories allow us to feel ourselves in the future, not just think ourselves there.

Next, fiction can deal with complexity. The future consequences of climate change tend to be imagined around keeping global heating below certain temperature increase targets: 2°C, 3°C, 4°C. Science can give us some sense of the climate consequences of each – but too often, that means that we imagine a binary future where a 1.5°C rise is safe and everything else is hell.

But it's not that simple, and that's where fiction comes in. The game allows players to create a world with a stable climate that's even more unjust than the one we live in now – or a fair world that's unlivably hot.

One stream takes us down an ultra-capitalist route, where corporate

→

visas control climate-driven migration and dictate global policy, and businesses exploit the fact that they own climate-ready technology. Such a future might end up being good in temperature terms, but raise challenges in terms of global equity. Another illustrates a future where a Global Climate Council becomes an authoritarian government, keeping warming to 1.5°C but crushing freedom in the process. Other choices build a healthier version of the world at that same temperature: where citizens are protected, democracy flourishes, and governments work together. This reminds us that there's not a single good future and a single bad one: instead, there's an infinite number of complex futures, and only we can control which we get.

ADVENTURE

Our game helps players understand that we can shape our future, not just witness it. For many, that feeling of power over climate change has been lost in recent years.

As climate activists saw inaction stall progress, many stopped hedging. They developed clear and unequivocal messages: if we don't stop climate change right now, there is no future for us on this planet. It got people to understand the scale of the threat. But it also made some people believe there's nothing they can do: this is the future of the planet and big international climate change meetings, they thought, something bigger than me, and where's the point in trying to change that?

→

Fiction – and especially a branching narrative style – puts control in the hands of the reader. It reminds them that the important decisions are the ones that lie ahead of us, not behind us.

To take action on climate change, we have to know what science says is likely to happen. But we must also know what it might be like to feel the heat of an unprecedented bushfire on our skin, the spike of fear when multiple crops fail, or the relief of breathing cleaner air in greener cities, halting sea ice loss in the Arctic, and the joys of living in more equitable societies.

Facts keep us realistic. But to really believe them, we have to be willing to deal in fiction.

- To see the full list of credits TURN TO PAGE 284.
- To start playing TURN TO PAGE 6.

Credits

CREATED BY

Sam Beckbessinger

Sam Beckbessinger is the author of a book called *Manage Your Money Like a Fucking Grownup* (Jonathan Ball and Little, Brown), which is sold in six countries and has been on the top 10 South African bestseller list for nearly two years now. She's also a cartoon scriptwriter, has published three picture books for young children, and wrote for Marvel's *Jessica Jones: Playing With Fire* on Realm. Her first novel, *Girls of Little Hope* (with Dale Halvorsen), is coming out in 2023. She gets way too excited about gross body fluids and still has a blog even though it's 2022. She lives in London. sambeckbessinger.com

Dr Simon Nicholson

Simon Nicholson is Associate Professor of International Relations in the School of International Service at American University in Washington DC. He also heads American University's Center for Environment, Community, and Equity and the Institute for Carbon Removal Law and Policy. Simon writes and teaches about global environmental politics and about what emerging technologies mean for environmental futures.

→

Dr Christopher Trisos
Christopher Trisos directs the Climate Risk Lab at the African Climate and Development Initiative, University of Cape Town. His research maps climate change risks to people and ecosystems, and combines insights from environmental and social sciences to understand how we can adapt to climate change risks. He is an author for the Intergovernmental Panel on Climate Change.

SHORT FICTION BY

Lauren Beukes
Lauren Beukes is the award-winning and internationally best-selling South African author of *The Shining Girls*, *Zoo City* and *Afterland*, among other works. Her novels have been published in 24 countries and are being adapted for film and TV. She's also a comics writer, screenwriter, journalist and documentary maker. laurenbeukes.com

Maria Turtschaninoff
Maria Turtschaninoff is a Finnish writer known for crafting lyrical, historically inspired fantasy stories starring strong female protagonists. She is the author of the Red Abbey Chronicles, a multiple award winning YA fantasy series sold to nearly 30 countries. Growing up in Finland means growing up close to nature, and both the forest and the sea are important inspirations for her work. www.mariaturtschaninoff.com

→

Rajat Chaudhuri

Rajat Chaudhuri is a bilingual author, environment columnist and climate activist. His works include novels, short story collections and translations. Chaudhuri is also editor of The Best Asian Speculative Fiction anthology and a co-editor of the Multispecies Cities-Solarpunk Urban Futures collection. His biopunk novel, The Butterfly Effect, was twice listed by Book Riot as a 'Fifty must read eco-disasters in fiction' and among 'Ten works of environmental literature from around the world'. Chaudhuri is a Charles Wallace Creative Writing Fellow and he speaks about climate-literature interface issues in venues at home and abroad. He lives and works in Calcutta. www.rajatchaudhuri.net

Sophia Al Maria

Sophia Al Maria is a Qatari-American artist, writer, and filmmaker. Her work has been exhibited at the Gwangju Biennale, the New Museum in New York, and the Architectural Association School of Architecture in London. Her writing has appeared in Harper's Magazine, Five Dials, Triple Canopy, and Bidoun.

OTHER CONTRIBUTORS

Primary sponsors

Climate Interactive, the FLAIR Fellowship Programme: a partnership between the African Academy of Sciences and the Royal Society funded

→

by the UK Government's Global Challenges Research Fund, and the National Socio-Environmental Synthesis Center (SESYNC) under funding received from the National Science Foundation DBI-1639145.

Web development, book design, and production
Electric Book Works (electricbookworks.com)

Illustrations
Annika Brandow (annikabrandow.com)

Visual design
Karen Lilje (www.hybridcreative.co.za)

Copyediting
Louis Greenberg (louisgreenberg.com)

Scientific contributors
- Aditi Mukherji
- Andrew Jones
- Brian Beckage
- Carrie Hritz
- Chris Mahony
- Colin Carlson
- Daniele Visioni

→

- David Keller
- David Morrow
- Glenn Moncrieff
- Holly Buck
- Jane Flegal
- Jay Fuhrman
- Juan B. Moreno-Cruz
- Katharine J. Mach
- Laura M. Pereira
- Leslie Paul Thiele
- Maggie Clifford
- Mariia Belaia
- Michael Thompson
- Nicholas Simpson
- Oliver Geden
- Olúfẹ́mi Táíwò
- Sara Metcalf
- Shuchi Talati
- Valentina Aquila

Did you earn a badge in the story?
Cut it out here, or get creative and make your own.

Track the economy, temperature and conflict with these folding gauges.
Cut along the dashed lines. Then, fold up when things get worse.
Fold down if they get better.

Economy

Fold up
↑

Temperature

Fold up
↑

Conflict

Fold up
↑

Spread a little glue here
to stick the last page only to this spot.

Disaster!	5°C	All-out bloodshed
Yikes!	4.5°C	Nowhere to hide
Oh no...	4°C	Run!
Hmmm...	3.5°C	Danger!
Not great	3°C	Scary
So-so	2.5°C	Tense
This is nice	2°C	Snippy
Sweet	1.5°C	Friendly
Rocking!	1°C	Peaceful

Lightning Source UK Ltd.
Milton Keynes UK
UKHW050358190522
403191UK00003B/79